Scanning Electron Microscopy and
X-Ray Microanalysis

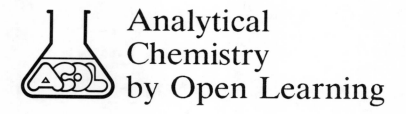

Analytical Chemistry by Open Learning

Project Director
BRIAN R CURRELL
Thames Polytechnic

Project Manager
JOHN W JAMES
Consultant

Project Advisors
ANTHONY D ASHMORE
Royal Society of Chemistry

DAVE W PARK
Consultant

Administrative Editor
NORMA CHADWICK
Thames Polytechnic

Editorial Board
NORMAN B CHAPMAN
*Emeritus Professor,
University of Hull*

BRIAN R CURRELL
Thames Polytechnic

ARTHUR M JAMES
*Emeritus Professor,
University of London*

DAVID KEALEY
Kingston Polytechnic

DAVID J MOWTHORPE
Sheffield City Polytechnic

ANTHONY C NORRIS
Portsmouth Polytechnic

F ELIZABETH PRICHARD
*Royal Holloway and Bedford
New College*

Titles in Series:

Samples and Standards
Sample Pretreatment
Classical Methods
Measurement, Statistics and Computation
Using Literature
Instrumentation
Chromatographic Separations
Gas Chromatography
High Performance Liquid Chromatography
Electrophoresis
Thin Layer Chromatography
Visible and Ultraviolet Spectroscopy
Fluorescence and Phosphorescence Spectroscopy
Infra Red Spectroscopy
Atomic Absorption and Emission Spectroscopy
Nuclear Magnetic Resonance Spectroscopy
X-ray Methods
Mass Spectrometry
Scanning Electron Microscopy and X-Ray Microanalysis
Principles of Electroanalytical Methods
Potentiometry and Ion Selective Electrodes
Polarography and Other Voltammetric Methods
Radiochemical Methods
Clinical Specimens
Diagnostic Enzymology
Quantitative Bioassay
Assessment and Control of Biochemical Methods
Thermal Methods
Microprocessor Applications

Scanning Electron Microscopy and X-Ray Microanalysis

Analytical Chemistry by Open Learning

Author:
GRAHAME LAWES
Royal Holloway and Bedford New College

Editor:
ARTHUR M. JAMES

on behalf of ACOL

Published on behalf of ACOL, Thames Polytechnic, London
by
JOHN WILEY & SONS
Chichester · New York · Brisbane · Toronto · Singapore

© Crown Copyright, 1987

Published by permission of the Controller of
Her Majesty's Stationery Office

All rights reserved.

No part of this book may be reproduced by any means, or
transmitted, or translated into a machine language without the
written permission of the publisher.

Library of Congress Cataloging in Publication Data:

Lawes, Grahame
  Scanning electron microscopy and X-ray microanalysis.
  (Analytical chemistry by open learning)
  1. Scanning electron microscope. 2. X-ray microanalysis.
  I. James, Arthur A. M. (Arthur M.),   1923–
  II. ACOL (Firm : London, England)   III. Title.   IV. Series.
  QH212.S3L38   1987   543'.0812   87-10458

  ISBN 0 471 91390 1
  ISBN 0 471 91391 X (pbk.)

British Library Cataloguing in Publication Data:

Lawes, Grahame.
  Scanning electron microscopy and X-ray microanalysis.—(Analytical
  chemistry).
  1. Scanning electron microscope
  I. Title II. James, Arthur M. III. ACOL IV. Series
  502'.8'25   QH212.S3

  ISBN 0 471 91390 1
  ISBN 0 471 91391 X (Pbk.)

Printed and bound in Great Britain

Analytical Chemistry

This series of texts is a result of an initiative by the Committee of Heads of Polytechnic Chemistry Departments in the United Kingdom. A project team based at Thames Polytechnic using funds available from the Manpower Services Commission 'Open Tech' Project has organised and managed the development of the material suitable for use by 'Distance Learners'. The contents of the various units have been identified, planned and written almost exclusively by groups of polytechnic staff, who are both expert in the subject area and are currently teaching in analytical chemistry.

The texts are for those interested in the basics of analytical chemistry and instrumental techniques who wish to study in a more flexible way than traditional institute attendance or to augment such attendance. A series of these units may be used by those undertaking courses leading to BTEC (levels IV and V), Royal Society of Chemistry (Certificates of Applied Chemistry) or other qualifications. The level is thus that of Senior Technician.

It is emphasised however that whilst the theoretical aspects of analytical chemistry can be studied in this way there is no substitute for the laboratory to learn the associated practical skills. In the U.K. there are nominated Polytechnics, Colleges and other Institutions who offer tutorial and practical support to achieve the practical objectives identified within each text. It is expected that many institutions worldwide will also provide such support.

The project will continue at Thames Polytechnic to support these 'Open Learning Texts', to continually refresh and update the material and to extend its coverage.

Further information about nominated support centres, the material or open learning techniques may be obtained from the project office at Thames Polytechnic, ACOL, Wellington St., Woolwich, London, SE18 6PF.

How to Use an Open Learning Text

Open learning texts are designed as a convenient and flexible way of studying for people who, for a variety of reasons cannot use conventional education courses. You will learn from this text the principles of one subject in Analytical Chemistry, but only by putting this knowledge into practice, under professional supervision, will you gain a full understanding of the analytical techniques described.

To achieve the full benefit from an open learning text you need to plan your place and time of study.

- Find the most suitable place to study where you can work without disturbance.

- If you have a tutor supervising your study discuss with him, or her, the date by which you should have completed this text.

- Some people study perfectly well in irregular bursts, however most students find that setting aside a certain number of hours each day is the most satisfactory method. It is for you to decide which pattern of study suits you best.

- If you decide to study for several hours at once, take short breaks of five or ten minutes every half hour or so. You will find that this method maintains a higher overall level of concentration.

Before you begin a detailed reading of the text, familiarise yourself with the general layout of the material. Have a look at the course contents list at the front of the book and flip through the pages to get a general impression of the way the subject is dealt with. You will find that there is space on the pages to make comments alongside the text as you study—your own notes for highlighting points that you feel are particularly important. Indicate in the margin the points you would like to discuss further with a tutor or fellow student. When you come to revise, these personal study notes will be very useful.

∏ When you find a paragraph in the text marked with a symbol such as is shown here, this is where you get involved. At this point you are directed to do things: draw graphs, answer questions, perform calculations, etc. Do make an attempt at these activities. If necessary cover the succeeding response with a piece of paper until you are ready to read on. This is an opportunity for you to learn by participating in the subject and although the text continues by discussing your response, there is no better way to learn than by working things out for yourself.

We have introduced self assessment questions (SAQ) at appropriate places in the text. These SAQs provide for you a way of finding out if you understand what you have just been studying. There is space on the page for your answer and for any comments you want to add after reading the author's response. You will find the author's response to each SAQ at the end of the text. Compare what you have written with the response provided and read the discussion and advice.

At intervals in the text you will find a Summary and List of Objectives. The Summary will emphasise the important points covered by the material you have just read and the Objectives will give you a checklist of tasks you should then be able to achieve.

You can revise the Unit, perhaps for a formal examination, by re-reading the Summary and the Objectives, and by working through some of the SAQs. This should quickly alert you to areas of the text that need further study.

At the end of the book you will find for reference lists of commonly used scientific symbols and values, units of measurement and also a periodic table.

Contents

Study Guide xiii

Supporting Practical Work xv

Bibliography xvii

1. SEM-Instrumentation 1
    1.1. Introduction 1
    1.2. Operating Principles 1
    1.3. Specimen/electron Interactions 11
    1.4. Detectors 16
    1.5. Operating Conditions and Limitations 22

2. Specimen Preparation 31
    2.1. Specimen Characteristics 31
    2.2. Drying Techniques 34
    2.3. Coating 42
    2.4. Cryo-SEM 52

3. SEM X-Ray Microanalysis – Instrumentation . . . 54
    3.1. Introduction 54
    3.2. X-Ray Production in the SEM 55
    3.3. Wavelength Dispersive Systems (WDS) 64
    3.4. Energy Dispersive Systems (EDX) 68
    3.5. Operating Conditions and Limitations 74
    3.6. Data Handling 76
    3.7. Other Surface Analytical Techniques 85

Self Assessment Questions and Responses 88

Units of Measurement 97

References 103

Study Guide

In this Unit we will begin by examining the operating principles of the scanning electron microscope (SEM), and discover that there is no light microscope equivalent. As we do so we will discover that the environment inside the SEM, imposes some restrictions on specimens. Most of these limitations can be overcome by careful specimen preparation, many of the proven techniques currently used are covered in the text.

When the SEM was first introduced, X-ray emissions from the specimen were of little practical use, now they can be put to practical use. We will look in detail at their production, detection and characterisation, and find out how they can provide information about the elemental composition, and distribution within the specimen. We will consider the practical limitations of present systems, and introduce ways of processing the data to provide quantitative information.

The aim throughout the Unit, has been to avoid the mathematics (which are well documented elsewhere), and concentrate on basic principles. Correction calculations are very complex, seldom undertaken without a computer and are continually being revised. Providing the user has an awareness of the problems, they can be left to the development and software engineers. X-ray analysers are self-contained instruments, for this reason the section dealing with them is also largely self-contained. Those with previous knowledge or experience of SEM, should find the final section useful by itself, with only occasional reference to the earlier text.

It is important to realise from the outset, that this Unit is not comprehensive, rather it forms a base on which to build, preferably with some practical experience.

Supporting Practical Work

1. GENERAL CONSIDERATIONS

Although facilities for practical scanning electron microscopy are becoming more widely available, relatively few of these instruments are provided with X-ray microanalysis. These facilities are more often available in laboratories other than chemistry, eg biology or geology. The first two experiments outlined below are intended for those with access to basic SEM, and the third experiment for those with access to X-ray microanalytical facilities.

2. AIMS

There are four principle aims:

(*a*) to provide basic experience in specimen processing;

(*b*) to acquaint students with basic operating principles of the scanning electron microscope;

(*c*) to illustrate the effect of various operating conditions on the resulting images;

(*d*) to show the advantages and disadvantages of X-ray microanalysis over other analytical techniques.

3. SUGGESTED EXPERIMENTS

(*a*) Prepare a delicate biological specimen (eg amoeba, blood sample), for examination in the SEM. Prepare one sample by simple air drying, and another by the recommended techniques of chemical fixation, dehydration and critical point of freeze drying. Compare the results and comment upon them.

(*b*) Take a series of electron photomicrographs from a specimen of your choice, using the SEM under different operating conditions (eg accelerating voltage, working distance etc). Comment on the results and explain the differences in appearance.

(*c*) Prepare a sample of glass fibre (loft insulation material) and mineral wool (cavity insulation). Photograph the two types of specimen and carry out X-ray microanalysis on both. Compare the resulting spectra and quantitative data (if obtainable) and comment upon them. Compare the results of the analyses with those obtained by an alternative technique of your choice (eg atomic absorption spectroscopy). Comment on the results, and discuss the merits of both techniques.

Bibliography

The following books on SEM and X-ray Microanalysis would be suitable for further reading and reference:

Goodhew, P.J. *Electron microscopy and analysis.* Wykeham Publications (1974).

Goldstein, J.I., Newbury, D.E., Echlin, P., Joy, D.C., Fiori, C. and Lifsin, E. *Scanning electron microscopy and X-ray microanalysis.* Plenum Press (1981).

Morgan, A.J., *X-ray microanalysis in electron microscopy for biologists.* Oxford University Press. Royal Microscopical Society, Oxford (1985).

Glauert, A.M. (ed.) *Principles and practice of electron microscope operation*, Vols 1–3. North-Holland Publishing (1974).

Hayat, M.A. (ed.) *Principles and techniques of scanning electron microscopy*, Vols 1–7. Van Nostrand Reinhold Co. (1980).

1. SEM-Instrumentation

1.1. INTRODUCTION

Since its invention in the early 1960's, the Scanning Electron Microscope has moved out of the specialist laboratory, and become an everyday tool, used by many. It opens up a world of amazing three-dimensional structures, which are easily interpreted, even by those with little experience.

As technology produces smaller and smaller structures, many too small to see with any conventional light microscope, the SEM has become increasingly essential and widespread. With more instruments being used, relative costs have fallen, and there is intense competition between manufacturers to develop more powerful and sophisticated instruments. Developments from the space and computer industries have been borrowed, and many modern instruments are digitally controlled and often have image enhancing systems built-in to 'clean up' poor images.

There can be little doubt that the future of the SEM has never looked brighter!

1.2. OPERATING PRINCIPLES

Although at first a scanning electron microscope may appear quite complex (Fig. 1.2a), we can simplify the instrument by separating it into three major sections:

electron-optical 'column';
vacuum system;
electronics and display system.

As we look at each of these you may soon realise that the instrument relies very much on the ingenious use of principles familiar to most students.

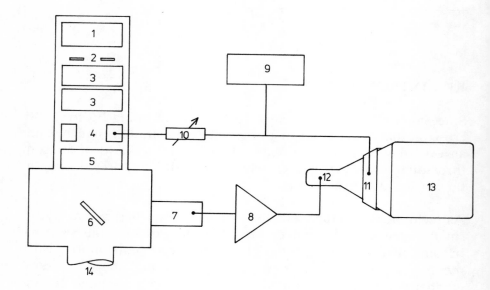

Fig. 1.2a. *Layout of Scanning Electron Microscope*

Electron optical column

1. Electron gun
2. Anode disc
3. Condenser lens
4. Scan coils
5. Objective lens
6. Specimen
7. Detector

Display/electronics

8. Signal amplifier
9. Waveform generator
10. Magnification control
11. Scan coils
12. CRT brightness control
13. CRT display screen
14. Vacuum Connection

Let's begin with the optical part of the instrument, the electron-optical 'column'. Unfortunately there is no light microscope equivalent to the sem, so we must look at each component of the column in some detail.

The first thing we need is a source of illumination. No visible photons of light for us, but an invisible beam of electrons. The beam is produced from an electron 'gun'. A cross-section through a simplified gun is shown in Fig. 1.2b.

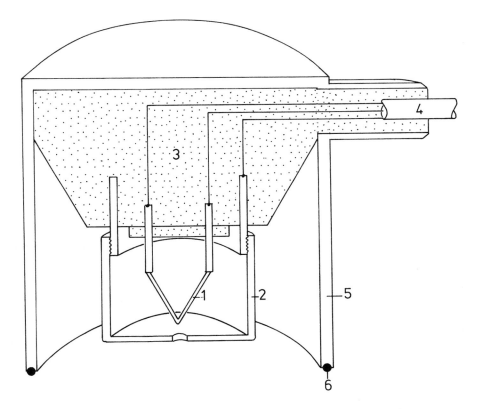

Fig. 1.2b. *Cross-section through an electron 'gun'*

1. Tungsten filament
2. Wehnelt cylinder
3. Insulation
4. Electrical supply
5. Column casing
6. Vacuum seal

A V-shaped tungsten filament (1) is heated electrically to about 2700 K. This high temperature causes many of the electrons in the tungsten to become sufficiently excited for them to escape. The process is called thermionic emission.

Once freed the electrons would be quickly recaptured by the filament, because in losing them, it will have become positively charged. Applying a high negative voltage (typically 2–25 kV) between the filament and a nearby earthed anode disc, accelerates the electrons away from the filament. Their velocity depends on the accelerating voltage and is only a fraction of the speed of light. Because of the high voltages applied to the gun, good electrical insulation (3) is essential!

Enclosing the filament in a metal cylinder (2), usually called the Wehnelt cylinder or cathode, shapes the beam electrostatically, so that it emerges 10–50 μm in diameter. Unfortunately in air, or any other atmosphere, the electrons would be scattered by collision with gas molecules. They could travel only a few millimetres. A vacuum system is connected (Fig. 1.2a, 14) to the column, so as to remove most of the gas molecules from the beam's path.

As we shall see the ultimate performance of the SEM is mainly limited by the diameter of the beam. In order to improve performance we must be able to control it. When we look in detail at operating conditions (Section 1.5), you will see that this is one of the most important operating parameters.

In Fig. 1.2a the upper two lenses, the condensers (3), control the beam's diameter. They demagnify it, reducing it from about 50 μm to around 5 nm. It seems odd but the only lenses used in an SEM are not used to magnify at all, in fact they do the exact opposite!

∏   You may not be very familiar with these small units of length – so try working out the number of times that the beam has been demagnified.

Hopefully you remembered that a nanometre is a thousandth of a micrometre. Dividing 50 μm (millionths of a metre) by 5 nm should have given you 10 000. You see it's rather like looking the *wrong* way through a very powerful telescope, the image of the beam is diminished ten thousand times!

Electron lenses are very different from their optical counterparts. Fig. 1.2c shows their main features.

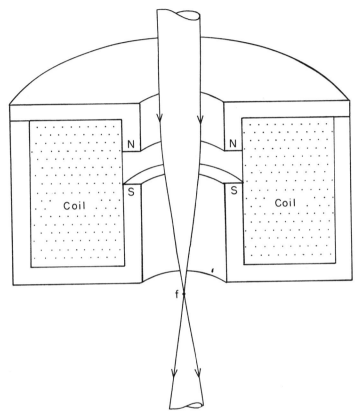

Fig. 1.2c. *Simplified electron lens (Cross-section)*

A coil of wire, with its axis aligned along the beam's path, is partially enclosed in a cylindrical iron case. There is a small gap in the inner bore of the case. When a direct current is passed through the coil an electromagnet is produced, with magnetic poles (N and S), created at the gap in the iron case. This is the 'polepiece gap', and it is really the magnetic lines of force, bridging this space, that form the lens.

Because of their charge, electrons are deflected as they intersect the lines of force. All electrons, entering the bore at the top of the lens, converge at a focal point (f), below the lens. A single point of focus is only produced if all the electrons have the same energy. This means the gun's accelerating voltage must be kept very stable, any variation will result in electrons of different wavelength or 'chromatic abberation'.

Changing the current through the coil changes the magnetic field strength. This in turn changes the angle through which the electrons are deflected, resulting in a change in the focal length of the lens.

Now that we have some basic components we can look again at the overall diagram (Fig. 1.2a). It shows a typical lens arrangement. Two condenser lenses (3) control the beam diameter, and a third lens, the objective (5), ensures that the beam has its smallest diameter when it strikes the specimen surface (6). This will focus our image.

When bombarded, many different interactions occur between the specimen and the electrons. In the next Section (1.3) we will be looking at some of these in more detail. For the moment we will keep things simple by saying that electrons are emitted from the specimen. These are collected by a detector (7) which converts them into a small electrical signal. Different types of detectors will be covered later (Section 1.4). As we shall see in the next section, this signal contains a variety of information about a *single* point on the specimen's surface. For example, a point on a smooth surface would reflect incident electrons well, producing a strong signal in a suitable detector. On the other hand, an adjacent point might be part of small depression or hole which might produce a very small signal.

To form an image of the specimen we need to sample a large number of points over an area. The beam is systematically moved, point-by-point along a line, and the reflected electron signal is collected. After completing a line of 1000 points, the beam is moved quickly back to the start of the line. The beam is then shifted down one line width before repeating its 'scan'. A thousand lines are scanned, and the beam is rapidly restored to its initial starting point.

One complete scan, consisting of one thousand lines, each of one thousand points, is called a 'frame' (Fig. 1.2d).

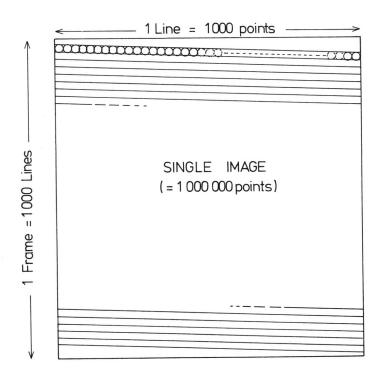

Fig. 1.2d. *Raster scan*

Your television at home works in much the same way, although in the UK, it only uses 625 lines per 'frame'. Technically this is known as a 'raster scan'. It is this scanning action that gives this type of electron microscope its name.

⊓   Can you guess how the beam is made to scan in this way?

It's our familiar electromagnets again, but this time they are not arranged along the axis of the beam but at right angles to it.

Two pairs of 'scan coils' are arranged at right angles to each other, and at right angles to the beam (Fig. 1.2a). The coils (4), are arranged in pairs (X and Y), on opposite sides of the column, between the objective and condenser lenses. Each pair is supplied with specially shaped current waveforms, produced in the waveform generator (9).

Controls allow us to scan the pencil of electrons over a chosen area of the specimen. We can use the signal collected from each point visited to construct an image.

A cathode-ray tube (CRT) is used to present the signal in an instantly recognisable form – a picture. Inside the CRT a second, independent beam of electrons is produced. When focused as a small spot on the tube's viewing screen (13), it appears as a small bright spot of light. In addition to the two pairs of scan coils located in the electron-optical column (Fig. 1.2a), another two pairs of coils (11) are fitted to the CRT. The same current waveform which produced the raster scan in the column, is also fed to these CRT scan coils. As a result the CRT beam and the column beam follow the same synchronised scanning pattern.

The signal from the detector (7) is fed to an amplifier (8) and on to the CRT's brightness control (12). A small signal reduces the spot's brightness and a large one increases it. Now that we have a complete imaging system let's just summarise the set-up.

The beam in the column looks at each point on the specimen in turn, if it is a reflective point a large signal is collected. The corresponding point on the CRT screen is made bright, and both the column beam and the CRT spot, move along the line to the next point. Again the signal is collected, the CRT screen brightness adjusted, according to the signal strength, and so on. After completing a line (1000 points) both beams 'fly back' to the start of the next line and continue. When the final point on the final (thousandth) line is reached, both beams return to their starting point. The scanning sequence is repeated all over again.

When the repetition is very rapid, say 25 times every second, the image on the CRT appears as a flicker-free TV picture. There are problems associated with very fast scan speeds, as we shall see later (Section 1.5).

So far we have managed to produce some sort of image of the specimen. If we are to use the imaging system as a microscope we must now provide magnification. Scanning electron microscopes enlarge the image in a unique and very simple way. Let us assume that the

Open Learning

beam in the column scans an area 20 cm square, and that the CRT screen is also 20 cm square. Our image is 'life-size', in other words there is a magnification of one (1×). Features on the specimen 1 cm apart will appear 1 cm apart on the viewing screen.

Now, for example, if we limit the area scanned in the column to 1 cm, but still display the image on the 20 cm screen, the image is magnified twenty times (20×).

SAQ 1.2a | It is quite common to use the SEM at a magnification of 40 000×, so assuming that we retain the 20 cm square CRT screen, work out the area being scanned in the column at this magnification.

SAQ 1.2b Draw a labelled sketch of an SEM and explain the important features.

Limiting the area scanned in the column is simple. All that is required is a means of reducing the current waveform sent from the waveform generator (9) to the column scan coils (4). This is shown (10) in Fig. 1.2a. A waveform of reduced amplitude deflects the column beam less and scans it over a smaller area. Remember, the current to the CRT scan coils is unaffected, so it continues to scan over the entire screen face.

A few other refinements are needed for a practical system. A comprehensive specimen 'stage' is fitted in the specimen chamber. This allows great freedom in orientating the specimen, so as to provide the required view. As well as three perpendicular axes of linear movement, X, Y, and Z (vertical), specimen rotation and tilt are essential.

Movable mechanical apertures are provided, to define the optical axis of the microscope (these will be dealt with later, Section 1.5).

It is usually necessary to provide an additional set of coils to correct an inherent defect in the optical system – astigmatism. Since this is a dynamic defect, constantly changing, it cannot be corrected totally by the manufacturer, and so needs an adjustable system for correcting it.

1.3. SPECIMEN/ELECTRON INTERACTIONS

In the previous section (1.2) we described the SEM's operating principles, saying only that a signal was produced when the electron beam scanned over the specimen. In this section we will look at some of the possible specimen/electron interactions more closely.

Another area from the previous section also needs some further discussion – the effective beam diameter. Although it is not uncommon to find SEM with a beam diameter of 5 nm, the diameter of the volume sampled, the so-called 'interaction volume', may be up to five hundred times larger. We will return to this later (Section 3.5). The effect is most apparent with 'bulk' samples (ie >1 μm^3). It is caused by electrons, and other resulting radiations scattering and diffusing through the sample, before emerging and being detected. We shall see that there are many possible interactions, Fig. 1.3a shows the links between the most important of these.

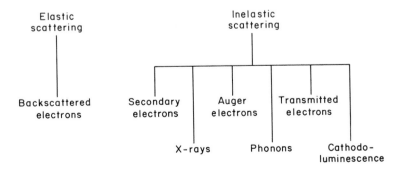

Fig. 1.3a. *Important specimen/electron interactions*

1.3.1. Unscattered Electrons

If the electrons have sufficient energy they may pass right through the specimen, suffering no effects at all. This is really a non-interaction, consequently the emerging electron beam contains no information about the specimen at all! For this to happen in the SEM, where the beam energy is typically 20–30 keV, specimens must be extremely thin (<1 μm).

1.3.2. Elastically Scattered Electrons

A beam electron (typically 20 keV), which passes close to a positively charged atomic nucleus, may be attracted by its opposite charge. As a result the electron changes its direction, with hardly any (<1 eV) loss of energy. The angle through which it is deflected (scatter angle), depends on how much energy it has, and how close it passes to the nucleus. It varies from almost zero to 180°, so the electron may be hardly affected at all, or may end up travelling back in the direction from which it came. The mathematics, which are unnecessary here, are well documented (Goldstein *et al*, 1981). In specimens of the same thickness, this so-called 'elastic' scattering, is more likely to occur in specimens of high atomic number (Z), and as you would expect, is also more likely if the incident electron has low energy.

Elastically scattered electrons are commonly used in SEM image formation. Many of those which are deflected through very large angles (>90°) re-emerge from the surface of the specimen, with high energy. When collected as a signal for imaging, they are usually called 'backscattered electrons' (bse), or reflected primary electrons. The intensity of the bse signal is dependent on the angle between the incident beam and the specimen surface. It follows then, that if the specimen has a rough surface, the signal changes with surface detail (topography). We mentioned earlier that the bse signal is dependent on the atomic number of the specimen. With flat 'polished' specimens, which may show little topographic detail, bse imaging provides very detailed 'atomic number contrast', elements of adjacent atomic number being discernable. In other words areas which may have identical surfaces, but are of different chemical compo-

Open Learning

sition, will have different signal intensities, and appear differently on the SEM screen. For example, chemical inclusions in glasses, ceramics and polished minerals show up well in bse imaging.

∏ Can you predict which of the following elements would produce the largest bse signal (and appear brightest on the SEM viewing screen): silver, lead, iron?

Assuming all had the same polished surface, the lead would appear brightest (largest bse signal), because its atomic number ($Z = 82$) is greatest. It would cause more elastic scattering than silver ($Z = 47$), which in turn would be brighter than iron ($Z = 26$).

1.3.3. Inelastic Scattering

Some of the incident primary electrons will interact with the orbiting shell electrons and atomic nuclei, losing a large proportion of their kinetic energy. These events are very complex, and there are a number of possible products, depending partly on how much energy is given up to the target atom (see Fig. 1.3a). Each of the possibilities will be discussed.

1.3.4. Phonon Production

Phonons are lattice oscillations set up in the specimen, as a result of electron bombardment. We need not consider them in any detail because, at present, they are of no use in SEM. Phonon production is only mentioned because it produces considerable heating of the specimen, and unless steps are taken to carry away the heat, some sensitive specimens may be permanently damaged (Section 2.3.1).

1.3.5. Secondary Electron Emission

Incident electrons may knock loosely bound conduction electrons out of the sample. If the weak electrons (<50 eV) are released within about 10 nm of the specimen surface, they may escape as low energy 'secondary' electrons (se). As with backscattered electrons the intensity of the se signal is dependent on the angle between the incident beam and the specimen surface, and like bse, is widely used in SEM topographical imaging.

1.3.6. Auger Electron Production

If an inner shell electron is knocked-out of its orbit by the incident primary beam, the atom must rearrange its outer shell electrons. If an electron from an outer shell drops down to an inner one, the atom is excited or ionised, and has excess energy. One way for the atom to lose this excess energy is for it to be transferred to a second outer shell electron, which can then be emitted from the atom. These are the Auger electrons. They have no use in SEM imaging, but have energies which are specific to the elements which produce them. Recently they have been used in specialised analytical instruments, Auger electron spectrometers (AES), to provide compositional information.

1.3.7. X-ray Production

Another way of filling the inner shell vacancies produced by inelastic scattering, is for an outer shell electron to drop down from an outer orbit, as in the previous case (Section 1.3.6). But instead of releasing the excess energy as an Auger electron, it can be released directly as a photon of electromagnetic radiation. If the amount of energy released is high, the photon will be in the X-ray part of the spectrum. In Section 3.2, when we look at X-ray production in more detail, we will see that there are two distinct forms of X-rays produced: 'characteristic X-rays' and the X-ray continuum or 'bremmstrahlung' (from German 'braking radiation').

1.3.8. Cathodoluminescence

Some specimens emit long-wavelength photons, in the visible or uv part of the spectrum, when exposed to the incident electron beam. The light emitted can be put to use in the SEM, though it tends to

Open Learning 15

be used to supplement information collected by other means (eg bse imaging). rather than on its own. Few materials have this property (Section 3.7), so we will not look at the process here. For detailed information see Goldstein *et al* (1981).

1.3.9. Electron Energy Loss

If the specimens are extremely thin, electrons may pass through them. While a few of these may pass through without any interaction at all (Section 1.3.1), most will have undergone inelastic scattering. When they emerge from the specimen they will have lost an amount of energy which is a characteristic of the elements present. At the moment SEMs do not make use of this information, since the specimens need to be extremely thin. We need not consider it further, except to say that transmission electron microscopes (TEMs), fitted with electron energy loss spectrometers (EELS), are now able to analyse the elements in specimens, using this process.

SAQ 1.3a List the important interactions that occur when an electron beam strikes the specimen and explain the conditions under which these occur. Which of the type(s) you have listed are of use in the SEM?

Summary

Elastically scattered electrons are those which change direction but maintain almost all of their energy (<1 eV loss). Elastic scattering is more likely to occur in atoms of high atomic number, and with low accelerating voltage beams. Elastic events are the most important in determining the shape of the 'interaction volume'.

Inelastic scattering causes the incident electrons to transfer a large proportion of their kinetic energy to the target atoms. It is more likely to occur in the lighter (low Z) elements. Inelastic scattering produces a wide range of useful information about the specimen surface and elemental composition.

One last point worth remembering – the overall efficiency of these specimen/electron interactions is very poor. Goodhew (1974) suggests that of all the incident beam energy absorbed by the sample, over 90% produces heat, about 2% produces X-rays, and the rest accounts for all the other radiative processes!

1.4. DETECTORS

In the previous section we discovered that there is a wide variety of specimen/electron interactions, we will now look at the ways in which these can be detected and put to use.

You will recall that if the samples are extremely thin, some electrons will pass right through them, emerging from the other side. These unscattered electrons, although easily detected, are totally unaffected by their encounter, contain no information about the sample and need not be considered further.

Elastically scattered electrons on the other hand contain useful information about the atomic number of the elements in the sample. They are frequently used in electron imaging particularly in materials and earth sciences.

There are two principal detectors for these high energy (>10 keV) electrons; the scintillator/photomultiplier type and the solid state

Open Learning

type. The first is rapidly being replaced by the more recent solid state device, which we will look at in detail here. However, the older scintillator detector is still the most widely used for secondary electron imaging, so we will be discussing it too, in some detail (Section 1.4.2).

1.4.1. Solid State Electron Detectors

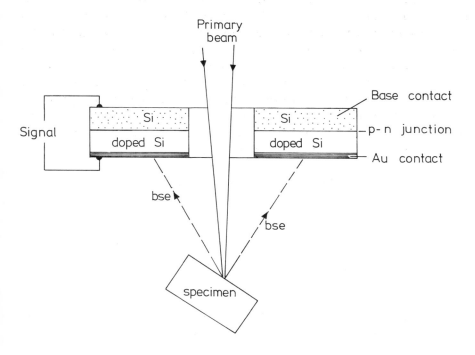

Fig. 1.4a. *Solid state electron detector*

These rely on the production of electron-hole pairs in a semiconducting wafer. The wafer, usually silicon, is several millimetres thick and in the form of a flat sheet, mounted onto a base contact (Fig. 1.4a). It is 'doped' with impurities to produce layers with different properties. In one type of silicon (usually referred to as n-type) the majority carriers are electrons, whereas in a differently 'doped' sample (so called p-type), the majority carriers are electron vacancies or 'holes'. Referring to any basic textbook on electronics will explain this more fully. In our detector the silicon wafer has a p-type layer and an n-type layer, forming a p–n junction. The upper surface has a very thin metal (usually gold) electrode evaporated onto the surface. The p–n junction means that there is an empty conduction band, separated from a filled valence band, by a band gap. When hit by the energetic backscattered electrons from our specimen, inelastic scattering occurs in the semiconductor. This moves electrons into the conduction band where they are free to move about. The resulting 'holes' left in the valence band can also move about. Normally these free electron/hole pairs would recombine, but if a small potential is applied across the wafer, they are swept apart resulting in a small electric current. In silicon about 3.8 eV of energy is needed to produce each electron/hole pair, so a single 10 keV electron scattered from our specimen would produce about 2600 electrons producing quite a small current! A very sensitive amplifier is needed to produce a signal large enough to vary the intensity of the SEM display screen.

In practical terms solid state detectors are very useful devices. Their main advantages are:

— they are insensitive to the low energy secondary electrons coming from the specimen. This is because electrons with less than 5 keV energy cannot penetrate the thin gold electrode and the inactive silicon layer;

— they are very small and robust so they can be located close to the specimen for maximum collection efficiency;

— they can be produced in a variety of shapes. A common type is in the form of a flat ring or 'annulus' mounted on the bottom of the SEM's final lens. This provides a hole for the beam to pass through, and allows collection over a large conical area.

Recently another variation of this has been developed in which the annulus is separated into four quadrants (Fig. 1.4b), each effectively a separate detector, and is particularly useful for mixing images to avoid high contrast and dark shadows in the image.

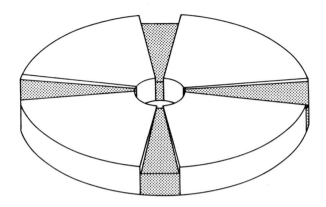

Fig. 1.4b. *Quadrant detector*

Unfortunately, these detectors do have one drawback – operating speed. Because of their own internal capacitance, present detectors do not respond very quickly, this means that they cannot be used at TV scanning rates.

Now we must consider the inelastically scattered electrons, and by far the most important are the low energy (<50 eV) secondaries. These contain a great deal of topographic information from the specimen and are the most widely used for routine specimen imaging.

1.4.2. Scintillator/photomultiplier Electron Detectors

These were the first and only detectors fitted to most early SEM's, and are still standard on most modern instruments.

Basically the detector works by converting the secondary electrons to photons, the photons to photo-electrons and finally back into electrons. It seems clumsy, but has remained little changed since its invention in the late nineteen fifties (Everhart and Thornley, 1960). Fig. 1.4c shows a typical arrangement, consisting of a wire grid or cage (C) in front of a scintillator tip (T) and, connecting this to the photomultiplier tube (PM), is a light pipe or guide (LG).

Most of the low energy secondary electrons from the specimen are attracted towards the end of the scintillator (T), because its rounded front face has a very thin (10–50 nm) layer of metal (usually aluminium) which has a positive potential (typically 10–15 keV) applied to it.

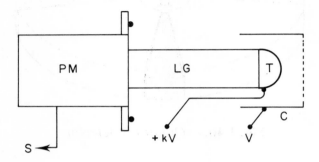

Fig. 1.4c. *Photomultipier/scintillator detector*

Collection of secondaries is not only 'line-of-sight', many of those leaving the specimen and travelling away from the detector will be bent around towards it. The secondary electrons having gained energy strike the scintillator tip and pass through the thin electrode. Once inside they are converted to visible light. This conversion occurs because the tip is made from plastic, glass or calcium fluoride which has been 'doped' with impurities such as europium. These materials convert the electrons to photons, ie they 'scintillate'. It is worth pointing out that the inorganic phosphors (eg zinc sulphide) commonly used in CRT tubes are unsuitable because, at the very low light levels involved here, they have a long 'after-glow' or persistence, so that the detector cannot be used at high scan speeds. Pawley (1974) carried out a comprehensive review of scintillators. New materials such as yttrium aluminium garnet (YAG) have recently been used, with considerable improvements in performance.

A light guide (LG), usually made of 'Perspex', transmits the signal (photons) by total internal reflection to the photomultiplier tube (PM). Generally the light guide passes through the vacuum seal of the SEM, so that the photomultiplier is outside the chamber, at atmospheric pressure. Here the light passes through the end window of the evacuated PM tube, striking the first in a series of electrodes. High electrical potentials are applied to these electrodes so that when hit by a single photon, many photo-electrons are emitted from the surface. Each of these is accelerated towards the next electrode where they cause numerous others photo-electrons to be freed. This is repeated through several more stages, causing a sort of electron avalanche. A single photon strike produces about 10^5 or 10^6 electrons at the output (S) of the PM tube. After further amplification the signal is fed to the SEM's display screen.

The wire cage (C) is really another electrode, connected to a supply (V), and can be used in two ways. Kept at about $+250$ eV it behaves like a Faraday cage, minimising the effect of the scintillator's potential on the SEM's scanning electron beam (reducing beam shift and astigmatism). At this potential it also improves secondary electron collection. If the cage is held at earth (0 eV) or made slightly negative (-50 eV) it no longer attracts secondaries but actually repels them, so that they do not contribute to the PM signal. However, those high energy backscattered (reflected) electrons which happen to be travelling in the right direction will be detected and amplified by the photomultiplier. Our detector has become a backscattered detector, and although not as efficient as the solid state ones described earlier (Section 1.4.1), is available at no extra cost!

Perhaps the major limitation of this type of detector is its large size. This makes it difficult to bring it very close to the specimen, so collection efficiency is poor (typically 50% for secondary and only 1–2% for backscattered electrons). Improvements in design and collection efficiency are being introduced (Robinson, 1980 and Wells, 1977).

∏   Can you think of any other advantages or disadvantages of these two types of detectors?

The PM/scintillator is faster than the solid state detector and can be used at TV scanning rates.

As yet, we have only briefly mentioned the other very important specimen/electron interaction, namely X-ray production. We will be looking at this in considerable detail, when we will discover that there are two distinct ways of detecting X-rays. Two distinctly different analytical instruments have evolved to exploit them. We will not consider them here, but will wait until we have dealt with X-ray production (Section 3.2 and Section 3.3).

Another of the electron/specimen interactions, cathodoluminescence, is much less commonly used for imaging, although it is an effect which is easily detected, using a photomultipier of the type discussed in Section 1.4.2. Unlike its use as an electron detector, it is used without a scintillator tip, photons being collected directly by the end of the light pipe, and amplified in the normal way.

1.5. OPERATING CONDITIONS AND LIMITATIONS

Now that we have covered the SEM's operating principles quite thoroughly, we ought to look at ways of achieving the best results from the system, and in doing so we will discover some of its limitations. It will soon become obvious that every time an improvement is achieved by changing one parameter, there is always a price to be paid, because one of the other operating conditions will deteriorate. It is important to realise that operating conditions are as variable as are the specimens we look at, and those suitable for one type of specimen may be totally unsuitable for another. Whilst there is no substitute for experience, we can establish a set of guidelines.

1.5.1. Resolution and Magnification

If you take the trouble to read any of the manufacturer's specifications, you will soon discover that the most important item in the data is the resolution. We shall see that this dictates the maximum magnification and ultimate power of the microscope.

What do we mean by resolution? Imagine two microscopically small particles separated from each other by a narrow gap (Fig. 1.5a). If when we view these through an optical system (microscope), we can see them as two distinctly separate bodies, then our microscope is capable of 'resolving' the distance between their centres (D). If we move the particles closer together we will eventually reach a stage where, instead of seeing two discrete and separate particles, we will see only a single, somewhat blurred shape, with the particles either apparently touching or overlapping (Fig. 1.5a). Our microscope is now incapable of resolving the distance (D) between them.

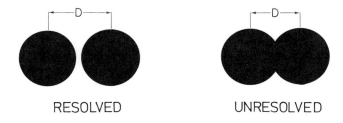

Fig. 1.5a. *Resolution*

In the SEM the resolution is mainly limited by the diameter of the probing beam of electrons, with which we scan the specimen ('spot size'). Fig. 1.5b shows the effect of beam diameter. A large diameter beam would produce an indistinct signal, from the two adjacent particles and might only just detect their presence. On the other hand, a smaller beam would actually produce a signal showing them as two separate particles.

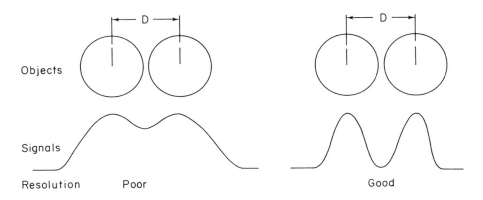

Fig. 1.5b. *Effect of spot size on resolution*

In SEMs the resolution is limited to about the same distance as the beam diameter (typically 5 nm). Manufacturers are continually struggling to produce SEMs with finer diameter beams, and better resolutions, but there are numerous problems, eg spherical, and chromatic aberrations (Goldstein, *et al* 1981), which make their progress difficult. Besides, we will see later when we look at signal/noise ratios (Section 1.5.4), that there is a price to pay for using very small spot sizes.

Magnification is closely linked to resolution. For most practical purposes the SEM's display screen is limited to about 20 cm square, and present technology only allows these displays to be scanned with about 2000 lines. This means that adjacent lines are separated by 0.1 mm, which incidentally, is about the resolution limit of the eye. Assuming a minimum spot size of 5 nm, we can calculate the maximum magnification by:

$$\text{magnification} = \frac{\text{screen resolution}}{\text{beam diameter (spot size)}}$$

ie magnification = 0.1 mm/5 nm = 20 000 times

It is quite common to find manufacturers providing much higher magnifications than those one would expect from their specified resolution capabilities. Although sometimes useful for focusing, and more comfortable for viewing, these over-magnified images contain no more information, because of the limit of resolution. They may be larger (and possibly easier to view), but they are not more detailed!

1.5.2. Accelerating Voltage

Accelerating voltage, often referred to simply as 'kV', plays a very important role in determining microscope performance. Few SEMs operate much above 30 keV maximum, some now offer voltages as low as a few hundred eV, and most are variable over this range.

Chromatic aberration is reduced at higher accelerating voltages, so for best resolution, the highest voltage should be used. The electron beam produced is obviously much more energetic. It can, and will, penetrate much more deeply into the surface of the specimen, particularly if it is of low atomic number (eg most biological materials). As an example, even in materials such as aluminium, where a 5 keV electron might only penetrate about 1 μm, a 30 keV electron could reach ten times this depth. The depth to which the electrons penetrate, and the resulting 'interaction' volume (Fig. 1.5c), is of considerable importance.

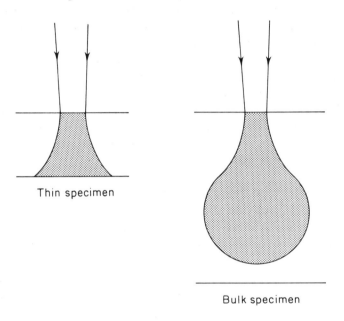

Fig. 1.5c. *Interaction volume*

This is not just an academic consideration, beam energy and interaction volume are important in quantitative X-ray microanalysis, as we shall see later (Section 3.5). Where images are concerned, these tend to lack contrast at high energy, because of secondary emission from within the bulk of the specimen. In extreme cases underlying detail can be seen, though not clearly, beneath the specimen surface, giving specimens a sort of semi-transparent appearance. Finally, because the sample is absorbing much more energy at high kV, specimen charging and damage is more common.

1.5.3. Scan Speed

The speed at which the electron beam scans over the specimen (and display) is variable over quite a range. At 'TV rate' for example the beam scans twenty-five complete frames (or pictures) every second. When recording photographs it is not uncommon to take one hundred seconds to complete a single frame. Fast scan speeds (TV rates in particular) mean that the specimen can be examined in 'real time', and because the displayed image is being updated many times a second, it makes it very much easier to orientate the specimen and locate areas of interest. Slow scan speeds are awkward to use, after making some adjustment to the specimen position for example, we may need to wait tens of seconds, before the screen image is updated to show the new orientation. Why then the need for slow scans at all? For photographic recording of the image, it doesn't matter if it takes several minutes to acquire a single picture, particularly if image quality is improved as a result. The explanation is simple – the problem is one of 'noise'.

At TV rates the electron beam is scanning about one thousand points (pixels) per line, about one thousand lines per picture 'frame', and twenty five complete frames every second. This means that the beam dwells at each point (pixel) for just 0.04 microseconds every 40 milliseconds. We have learnt (Section 1.3 and 1.4) that the efficiency of the specimen/electron interactions and our detectors is quite low, so the actual signal derived from each point is very small indeed! Consequently, very high gain amplifiers are required, to boost the signal to a level, where it can be used to drive the SEM display screen. Such amplifiers tend to pick up random as well as signal electrons. These too, are processed indiscriminately, and result in electronic 'noise' (named after its effect on audio amplifiers). Here, we are dealing with video, so it appears as random light points superimposed on the SEM screen, and in extreme cases the noise can swamp the signal, so that its like viewing through a 'snowstorm'. Images of this type are said to have poor 'signal to noise' ratios.

1.5.4. Signal/noise (S/N) Ratio

One of the causes of poor signal/noise (S/N) ratios is the use of fast scanning speeds. If we leave the beam at every point on the specimen for a longer time, we will collect a larger signal from that point. The 'noise' inherent in the system is unchanged, so the signal to noise ratio is improved. Now less amplification is required, so that the signal swamps the noise, resulting in a clearer (less 'noisy') image.

1.5.5. Spot Size

Although resolution is directly proportional to spot size (beam diameter), operating with the smallest available spot size, although providing the highest possible resolution, also produces very 'noisy' images.

Where the high resolution is not required (eg at low magnifications), but fast scan speeds are helpful, then increasing the spot size will greatly improve the S/N ratio, and image quality. Bombarding the specimen with larger numbers of primary electrons will, predictably, produce larger signals in any of the detectors, because of the increased yield of secondaries, X-rays etc. The penalty in this case is that the specimen absorbs much more energy, and is therefore more likely to charge up and 'cook'.

1.5.6. Working Distance

The 'working distance' of the SEM is the distance from the final lens of the microscope, to the surface of the specimen. It can be varied, from about 5–40 mm, by moving the sample vertically up or down, using the specimen stage Z-axis shift. Modern instruments have a digital read-out of this distance (in mm), calculated automatically, from the final lens (focus) current.

Long working distances provide images in which surfaces close to the final lens are in focus, as well as those surfaces distant from it. The range over which near and far objects can be focused, whilst remaining acceptably sharp, is called the depth of focus, and will be familiar to anyone interested in photography. Long working distances give good depth of focus, short working distances give poor depth of focus. Resolution is proportional to the depth of focus, ie short working distances give increased resolution, and long distances give poorer resolution.

In practice, it is obviously a case of deciding whether you want to take low magnification pictures, where good depth of focus is more important than resolution, or *vice versa*.

1.5.7. Aperture Size

In Section 1.2 we mentioned that most microscopes are provided with mechanical apertures, which can be altered by the operator. These are small metal plates with very small (25–1000 μm diameter) holes (apertures) through them. They define the effective diameter of the lenses, and reduce some of the inherent aberrations. As far as microscope performance is concerned, large apertures give better resolution and better signal to noise ratios, but at the expense of depth of focus, and *vice versa*.

1.5.8. Beam Current

This is a direct measure of the number of electrons making up the beam, and is seldom greater than 100 μA. On an increasing number of instruments it is controlled automatically, if provision is made for operator control (by altering the gun operating conditions), it can be useful, particularly when examining beam sensitive specimens. Reducing the beam current, reduces charging and thermal damage to the specimen. Of course bombarding the specimen with fewer primary electrons does produce lower signals from it, and inevitably poorer signal to noise ratios.

SAQ 1.5a Define the term resolution, explain how magnification is dependent upon it, and list the operating conditions necessary to achieve maximum resolution.

Open Learning 29

SAQ 1.5a

Summary

Our discussion may have left you rather bewildered – with so many conflicting conditions and effects, where should you begin? Well, here are a few hints:

— the specimen may be damaged if you start by using high accelerating voltages, high beam currents and/or large spot sizes;

— remember that ultimate resolution is *not* always the goal;

— it will inevitably have to be a compromise;

— experimentation will lead to a better understanding of the problems, and invariably result in improved imaging.

Objectives

You should now be able to:

- describe the overall layout of an SEM and explain the function of each of the main components;

- list the various electron/specimen interactions and explain their importance in SEM;

- explain the construction and operation of the main types of electron detectors;

- give an account of the operating conditions and limitations of use of an SEM;

- define the terms resolution and magnification and explain their relationship.

2. Specimen Preparation

There are few objects and materials which cannot be examined in the SEM. Nowadays it is possible to look at materials ranging from large turbine blades to frozen mayonnaise. In this Part we will be considering some of the techniques which have been developed to overcome the problems posed by some of the more difficult materials.

As far as specimen size is concerned, it is only the dimensions of the specimen chamber that impose any limitations.

We will see that the effort involved in preparing a specimen can vary from none at all, in the case of metal samples, to several days of tedious dehydration, drying and coating in the case of some biological specimens.

2.1. SPECIMEN CHARACTERISTICS

If asked, any SEM user will quickly tell you that there is no such thing as an ideal specimen except, perhaps, a pure metallic silver test grid, or gold evaporated onto graphite. Anything else, they will tell you, will have its problems. Many of these problems can be easily overcome by suitable preparation techniques.

One way of looking at specimens and their characteristics is to consider them in relation to the rather hostile conditions inside the SEM.

2.1.1. Volatile Materials

You will remember that, in order to produce a beam, the microscope must operate with a high vacuum inside it. Any volatile material, in or on specimens will be vaporised at these low pressures. This is one of the most severe restrictions on the type of specimen that can be examined.

In extreme cases some chemical samples may be totally volatile at these low pressures! It is good practice, in the case of unknown samples, to expose them to a similar vacuum (eg in a freeze-drier or coating unit), before risking contamination of the microscope. Not surprisingly this property of specimens is usually referred to as 'outgassing', and can cause several problems. The vacuum becomes poorer, and can lead to a general loss of resolution. In addition some of the beam electrons collide with the freed vapour molecules causing beam instability. One of the worst culprits is also the most common – water. Water vapour from hydrated or partially hydrated samples is very undesirable, vacuum pumps have great difficulty removing it, and any residual molecules may be ionised near the high voltage gun. The very reactive hydroxyl ions resulting from this ionisation, can damage the gun's hot filament.

Outgassing can make the microscope very slow, or even impossible, to pump-down to a working vacuum.

As if this were not enough, movement of volatile materials can cause re-location, re-distribution and loss of elements within the specimens, making quantitative microanalysis, at best, inaccurate.

To overcome these problems, whenever possible, all specimens should be free from all volatile materials. Water is particularly undesirable, so specimens must be dry. Recently, very low-temperature (< -100 °C) specimen stages have been developed, Cryo-SEM, to try to overcome some of the limitations of the normal SEM. Cryo-SEM is described in Section 2.4.

2.1.2. Heat-sensitivity

We learnt earlier that much of the electron beam's energy is converted to heat, when it is absorbed by the specimen. This can produce quite dramatic localised increases in temperature, enough to damage delicate structures with low melting points. Varying some of the SEM operating conditions (eg spot size), will reduce the energy actually absorbed. Alternatively much of the heat can be carried away by ensuring that there is good contact between the sample and the microscope, by using conductive adhesives for instance. Coating the surface of the specimen with a thin metal layer (Section 2.3) will also help conduct the heat away. Unfortunately, in the case of very sensitive specimens (eg some low melting-point waxes), the only real solution is to use a cooled specimen stage, or a cryo-SEM.

2.1.3. Electrical Charging

Many, if not most, specimens are electrically non-conductive. Unless actually metal they have an electrical resistance which invariably leads to SEM-imaging problems.

As the electron beam sweeps over the specimen a surplus of electrons collects on its surface, unless the electrical resistance is low. This means that the specimen becomes more negatively charged, the longer the specimen is scanned. As the charge builds up it repels or deflects the approaching beam, causing image defects.

2.1.4. Electron Beam Sensitivity

A few materials, such as photographic emulsions and resists and some electronic semi-conducting devices are actually sensitive to electrons and may be permanently damaged when bombarded by them. These present serious problems, which can only really be overcome by varying the SEM operating conditions, for example using very low accelerating voltages (<1 keV), and as a result sacrificing resolution.

2.2. DRYING TECHNIQUES

2.2.1. Introduction

In an earlier section (2.1) we established that many specimens are initially hydrated to a greater or lesser extent. Biological samples in particular contain a very large percentage of water and other volatile materials, and these need very carefully controlled drying to avoid serious damage.

2.2.2. Fixation

It is common knowledge that biological material decomposes, as anyone with a broken refrigerator soon realises! At the microscopic level, this decay occurs very quickly due to the effect of enzymes within the cells (autolysis), and as a result of bacterial and fungal action.

Biological specimens must usually be exposed to chemical preservatives or 'fixatives' as soon as possible if life-like structures are to be observed. There are hundreds of different 'recipes' for fixatives, though most rely on a combination of buffered solutions of aldehydes and osmium tetroxide, to cross-link proteins etc inside the specimens. By converting the rather watery cell contents into a more stable gel, the internal and external structures are preserved, some structures hardened against drying damage, and chemical breakdown is stopped. It is quite beyond the scope of this work to go into any great detail on this subject, which is covered extensively elsewhere (Glauert, 1974; Hayat, 1980). All we will say is that usually the specimens are exposed to a suitable fixative whilst alive, if detail is to be faithfully preserved.

2.2.3. Air-drying

Let's consider a biological sample for a moment. Imagine a fresh, juicy plum, and what happens when it is allowed to air-dry – the result, a prune! This is a rather silly example, but it does show that at this scale, the specimen shape, size and surface detail is very dramatically altered, just by removing the water from it. Internally, chemical changes will also have occurred, and many of the elements will have been relocated (eg by diffusion), as water is removed. Now the skin of a plum is quite thick and relatively tough, but at the microscopic level, cell and tissue structures are much more delicate, and the damage to these will have been even more severe. Common drying artefacts include shrinkage, collapse, twisting, tearing and wrinkling. Clearly, if we want a realistic view of specimens we need an alternative way of drying them.

Several effects are at work here but, by far the most damaging is that of surface tension or, more accurately, interfacial tension. It results from the boundary between the gas (or vapour) and liquid phase passing through the specimen. When this boundary moves through the specimen, during drying or evaporation, a tremendous force is exerted at right angles to its surface. It has a profound flattening effect. The problem then is to get the specimen out of the liquid (if it is immersed) and the liquid out of the specimen, without allowing the liquid surface to pass through the specimen. At first this might seem, at best difficult, and at worst impossible. Fortunately it is neither, but to understand the methods, we need to do a little revision. Fig 2.2a shows the phase diagram for a compound. It shows the relationship between pressure, temperature and state. Remember, each state of matter – liquid, gas or solid can only exist within certain temperature and pressure ranges. There are two important points on the graph, the 'triple point' (T) and the 'critical point' (C). At the triple point, all three states of matter (gas or vapour, liquid and solid) exist at equilibrium and simultaneously.

On the diagram, the line between the triple and critical points represents the equilibrium between the liquid and the gaseous states. In reality it is the visible surface of the liquid, with liquid beneath and gas above. It is this boundary or surface that we must not allow the specimens to cross, or damage will occur.

Π   Do you know the Triple Point of water?

Atmospheric pressure (101 325 Nm^{-2}) and 273.15 K. At this tem-

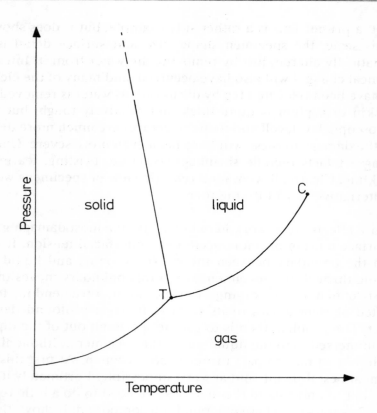

Fig. 2.2a. *Typical phase diagram*

perature and pressure, ice melts in the presence of water vapour, so all three states exist simultaneously.

The critical point (C) is the point at which the gas and liquid have exactly the same density. Above the critical temperature and pressure there is no distinct boundary between gas and liquid. This will be explained in more detail when we look at critical point drying (Section 2.2.5).

2.2.4. Freeze-drying

Historically this technique was the first used to overcome drying artefacts. The hydrated specimens are frozen as quickly as possible, typically with a cooling rate of hundreds of degrees Celsius per second. The size of the ice crystals formed inside the specimens depends upon the cooling rate, slow cooling results in large ice crystals and *vice versa*. If large crystals are allowed to form these may rupture cell membranes in biological samples, causing the collapse of cells and organelles, and deformation of the whole specimen. Extensive research has been carried out on ways of achieving very fast freezing, and we will not concern ourselves too much with the details. Generally, most methods rely on the use of very small specimens (<1 mm) and liquified nitrogen or helium (at −196 and −269 °C respectively).

The basic outline of a freeze-drier is shown diagramatically in Fig. 2.2b.

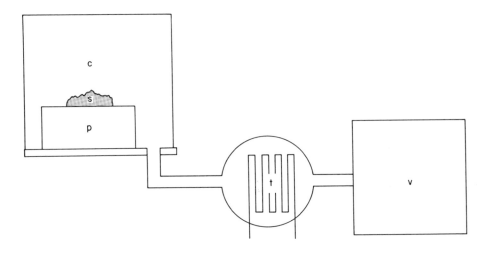

Fig. 2.2b. *Basic freeze-drier*

Once frozen, the sample (s) is transferred to the vacuum chamber (c), and is mounted on an electrically cooled platform or stage (p). Mechanical vacuum pumps (v) are then used to create a partial vacuum inside the chamber. By carefully controlling the temperature (usually between −40 and −70 °C) and pressure (typically 0.1 torr) the water in the specimens can be sublimed from the solid to the gas without a liquid state being present. Once vaporised, the water is sucked out of the chamber by the vacuum pump, and collected chemically on a dessicant (eg phosphorus pentoxide) or by a refrigerated condenser in the trap (t). When all the water has been sublimed off, the temperature and pressure can be restored to normal and the specimen will have been dried.

To understand why the specimen is undamaged we need to look again at a phase diagram (Fig. 2.2c). Initially the specimen contents are liquid, and are represented by point 1, in the liquid phase. Freezing converts the liquid to solid, the specimen moves to point 2. Although the temperature has been reduced the pressure is unchanged. Now the frozen specimen is transferred to the vacuum chamber, where the pressure is reduced (point 3), whilst the low temperature is maintained. Sublimation occurs, and all the water is

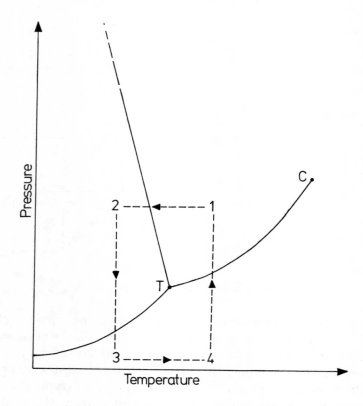

Fig. 2.2c. *Freeze-drying principle*

converted from solid to gas. Once the water has been removed the specimen can be allowed to warm-up to normal room temperature (point 4) and finally returned to atmospheric pressure (point 1). Now the specimen is completely dry, the water vapour having been removed and trapped by the equipment. This route has avoided crossing the damaging gas/liquid boundary with its damaging surface tension forces.

In addition to the problem of getting very fast cooling rates, the actual sublimation time is long, and is dependent upon specimen size and shape, temperature and, to a lesser extent, pressure. Another drawback with this technique is that it is difficult to determine when the specimen is completely dried, extended drying periods or trial and error seem to be the only solutions.

2.2.5. Critical Point Drying (cpd)

Although this process achieves the same end result as freeze-drying, it does it in the opposite way. Instead of using reduced temperature and pressure this method relies on increasing them.

Again it will help our understanding if we follow the process on a phase diagram (Fig. 2.2d).

As before the specimen starts, initially, in the liquid phase (point 1). This time the temperature is raised and, because the vessel containing the specimens is sealed, the pressure rises. When the critical temperature and pressure of the liquid are exceeded (point 2) the liquid gas boundary disappears, leaving the specimen dry.

This probably needs further explanation. As the temperature inside the vessel begins to rise, the density of the liquid phase decreases. The increasing pressure inside the vessel causes the density of the gas phase to increase. As we get closer to the critical temperature and pressure so the density of the gas and liquid get closer to each other. Eventually, when the critical point is exceeded gas and liquid have the same density, so there is no boundary to separate them, and they diffuse into each other.

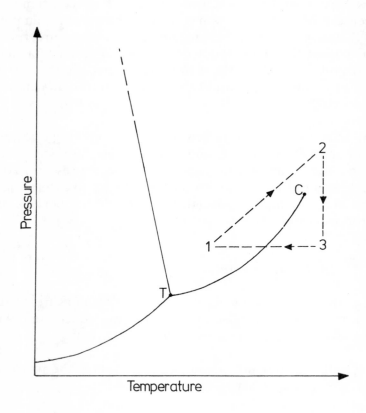

Fig. 2.2d. *Critical point drying principle*

Providing the temperature is kept above the critical value, the pressure can be lowered back to normal atmospheric (point 3), by slowly venting the gas from the vessel. Finally the vessel can be allowed to cool to normal ambient (point 1), and because the gas has been removed from the vessel it cannot reform as a liquid, so the specimen is left dry.

The liquid used in cpd is usually called the 'transitional fluid' and until now we have assumed that this is water, but unfortunately as Fig. 2.2e shows, its critical values are so high that any biological specimen would be 'pressure-cooked'!

Fluid	Formula	Critical Temp. °C	Critical Pressure (Atm)
water	H_2O	374.0	217.7
carbon dioxide	CO_2	31.1	72.9
nitrous oxide	N_2O	36.5	71.7
freon 13	$CClF_3$	28.9	38.2

Fig. 2.2e. *Critical values of useful transitional fluids*

Of the compounds shown, carbon dioxide is by far the most commonly used, having a critical temperature which does not damage biological specimens, and a critical pressure which is reasonably easy to contain. In addition it is cheap and readily available in cylinders. Unfortunately it is not miscible with water. This means that specimens containing water must first have this replaced by an intermediate liquid which is miscible both with the water and liquid carbon dioxide. Dehydration, the term used for this stage, is just a matter of immersing the specimens for a few minutes in solutions of increasing concentrations of ethanol or acetone (propanone), (say 30%, 50%, 70%, 90%, 100%). In this way the water is gradually replaced with the solvent.

There is a wide variety of commercial instruments available, each with its own particular operating and design features, but all rely on the same operating principle.

This technique is by far the most widely used for preparing hydrated specimens, mainly because of its speed. Most specimens, once in 100% solvent, can be processed in one or two hours.

SAQ 2.2a Explain the principles of freeze-drying and critical point drying. Discuss the relative advantages and disadvantages of the two methods.

SAQ 2.2a

Summary

To see realistic images of delicate, hydrated specimens it is not enough just to allow them to air-dry. We have only discussed the two most important and widely used methods, there are many more (eg freeze-substitution).

2.3. COATING

After drying the specimens are fixed onto small holders, usually called 'stubs', which resemble large drawing pins. A variety of adhesives are available, those with which do not outgas, have reasonable strength and good conductivity are best. Practically, epoxy resins, colloidal silver cements (or 'DAG'), and a sort of double sided tape are commonly used.

2.3.1. Specimen Charging

When non-conductive specimens are examined in the SEM they tend to collect a surplus of electrons on their surfaces, in other words they become negatively charged. In practice, unless actually metallic, almost all specimens can be affected. Those which have pointed projections (eg needle-shaped crystals) are particularly susceptible, because the charge accumulates at these points. Once the electrons start to build-up the image quality is badly affected. Intense contrast, dark 'halos' with brightly glowing centres and horizontal streaks, are common and obvious signs of charging. Occasionally, it causes a slow drift, and a sudden jumping-back, of the image. There are several effects at work here. Sometimes the negative charge deflects, or repels the electron beam so that in extreme cases it can build up sufficiently to slow down the incident primary beam, and the specimen actually becomes an electron mirror. Ocassionally some of the surplus electrons may eventually escape from the specimen, often after a considerable delay, causing a spurious signal in the detector. Specimen charging becomes more severe when high accelerating voltages are used. One way to overcome the problem would be to use a lower energy beam, and sometimes this is the only option available. However, when we discussed operating principles (Section 1.2), we found that a high energy beam is essential, if we are to achieve the resolution, necessary for high magnifications. Alternatively, using a very low beam current would help, but this would produce a very 'noisy' picture, and as we shall see later, makes microanalysis more difficult.

There is another way of approaching the problem. Instead of changing the operating conditions of the microscope, why not change the condition of the specimen, by making it conductive?

Spraying the specimens with commercial anti-static agents such as polyamine-derivatives, does little to improve conductivity in SEM specimens.

Coating the specimen surface with a very thin, uniform conductive layer does, however, provide a really effective solution to the problem. Metal coatings are the most obvious choice, and are the most widely used. The applied metal layer provides a low resistance path (to earth), for the electrons, and there is an added bonus, because the coating also conducts beam-induced heat away from the specimens. When the elemental composition of the specimens is also required (ie X-ray microanalyis), metal coatings can cause confusion and errors, and we will discuss this in detail in a later section. In these circumstances, a layer of carbon is often applied instead of metal, and although a rather poor conductor, it is a good secondary electron emitter, and is certainly better than no coating at all.

Careful consideration must be given to the thickness of the coating used, and there are conflicting requirements. On one hand, thick coatings are more efficient conductors, but if the layer becomes too thick, it may obscure very fine surface detail, and can interfere with microanalysis. It's rather like a thick blanket of snow that often makes it difficult to find the kerb!

Don't be afraid to try looking at the specimens, uncoated, initially. Unless, like some polymers and semi-conducting devices, they are very beam-sensitive, they are unlikely to be permanently damaged. If charging occurs, use a less energetic beam (ie reduced kV), faster scan speeds or backscatterd electron imaging, and if the charging continues, coat them.

Generally, it is better to apply a thin coating first, examine the specimen in the SEM, and repeat the coating if necessary. As a starting point try using a coating which is the same thickness as the maximum resolution of the microscope. For example, if the SEM is capable of resolving 7 nm, then initially try using a coating of about 7 nm. Unfortunately it is not practical to deposit coatings <10 nm thick by sputtering, only by vacuum evaporation. For general observation work, coatings of 15–25 nm are satisfactory.

2.3.2. Coating Materials

The material used for coating should have the following properties:

— good conduction (of heat and electrons);

— chemical inertness (resistant to oxidation and tarnishing layers);

— good secondary electron emmission (large signal);

— smallest possible grain size (to avoid 'decorating' the specimen);

— easy to produce thin films (by normal laboratory techniques).

Not many materials have all these properties; those most commonly used include gold, gold/palladium alloy and carbon.

Although there are quite a number of different coating techniques, many have been developed for very specific applications. We will only look at variations of two different techniques; namely vacuum evaporation and sputtering.

There is a variety of commercial equipment available, some capable of both techniques, and others dedicated to one or the other.

2.3.3. Vacuum Evaporation

Vacuum evaporation is the simpler of the two techniques, and can be used to produce a wide variety of metal coatings, and unlike sputtering, it can also be used for depositing carbon. The essential parts of a vacuum coating unit are shown in Fig. 2.3a.

The process is carried out inside a vacuum chamber (VC). Inside there are two terminals (T), connected to a low voltage, high current supply. A filament of tungsten wire (W) is stretched between these terminals, and will act as a heating element or source. A small amount of the metal to be evaporated, in the form of a fine wire, is suspended from the tungsten element. Pumps (P) are provided to remove the air from the chamber, and there is a vacuum gauge (VG), to measure the pressure in the system. Dry, mounted specimens (S) are placed at the bottom of the vacuum chamber, about 10 cm below the heater. The chamber is then sealed and evacuated. When a suitable vacuum has been reached (typically <0.1 torr) the current to the heater is switched on, and steadily increased. As the heater temperature is raised the fine metal wire melts and minute particles are driven off. These radiate outwards in straight lines, coating any surface directly exposed to them. This type of high vacuum coating

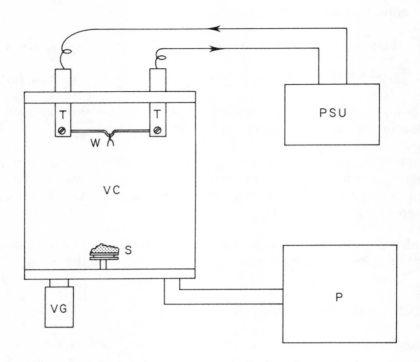

Fig. 2.3a. *Vacuum evaporation unit*

is 'line-of-sight' only, consequently, surfaces not directly facing the source of particles would remain uncoated. Coatings produced in this way are far from ideal, but can be improved upon.

Using several sources helps overcome the very directional nature of this technique and by mounting the specimens on a tilting and rotating platform, we can produce a much more uniform overall coating. Unfortunately, mechanical platforms, which must be rotated at about 10 rpm, have a tendency to throw specimens off their stubs. Determining the thickness of the deposit is possible using expensive commercial thickness monitoring instruments, which provide a direct read-out (usually in nm). Where this accuracy is not essential, fairly reproducible results can be obtained by evaporating a known weight (or length) of wire and positioning the specimens at a measured distance from it.

Open Learning 47

Although still used, particularly for very thin metal coatings (<10 nm), vacuum evaporation is largely being replaced by sputtering. Since carbon cannot, at present, be deposited by sputtering, only by vacuum evaporation, we will look at vacuum evaporation of carbon before moving on.

The same equipment can be used but now the wire heater is replaced by either a pair of carbon rods (held in contact with each other by springs), or a few strands of carbon fibre. Once the required vacuum has been attained, the carbon is heated, again using a high current (low voltage) supply. Carbon is evaporated and, as before, will coat anything inside the workchamber, including the specimens. Carbon is much less directional than metal when evaporated, so there is less need to move the specimen around during the coating. If a thickness monitor is not available then the evaporation time can be used for reproducible results.

It is advisable to use spectroscopically pure carbon, especially if the elemental composition of the specimens is to be determined by microanalyisis.

2.3.4. Sputter Coating

'Sputtering' or (diode sputtering) is the term used in physics for the process in which gas plasma is used to dislodge atoms from a metal cathode or 'target'. The equipment used is quite simple as you can see in Fig. 2.3b. Again the process is carried out in a glass chamber, in which a vacuum can be produced, by a pump. Attached to the lid of the chamber is an insulated metal plate, the 'target' (T), which is made from the coating material (eg gold/palladium alloy). There is a negative, high voltage low current DC supply (typically -2.5 kV at 20 mA) connected between the target and the earthed baseplate (or anode). The chamber is also provided with argon gas, supplied at low pressure, through a needle-valve.

Specimens (S) to be coated are attached to the baseplate and the pump is used to reduce the pressure. Low pressure argon is then leaked into the chamber, and maintained at about 0.1 torr by the vacuum pump, which is kept running. When the high voltage is

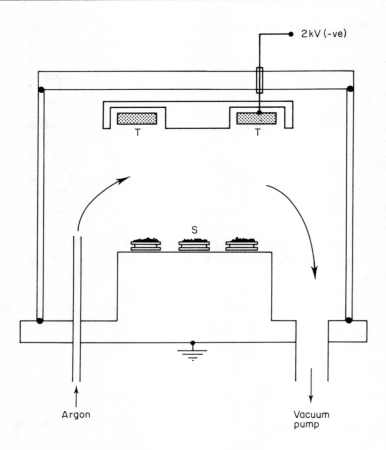

Fig. 2.3b. *Sputter-coating unit*

applied to the target much of the argon is ionised. The resulting electron/ion pairs move in opposite directions. The electrons travel downwards towards the positive baseplate, whilst the positively charged argon ions accelerate towards the negatvely charged metal target. On collision, many of the ions have sufficient energy to break the bonds between neighbouring atoms in the target. These dislodged atoms suffer repeated collisions with the migrating gas molecules, ions and electrons. Because of the turmoil inside the chamber the metal atoms take on a random motion, and behave rather like a metal 'smoke', settling on contact with any surface inside the chamber – including the specimens.

Regulating the supply of argon into the system controls the ionisation current (shown on a milliammeter), and provided other operating conditions (eg voltage) are kept constant, determines the deposition rate per unit time.

Commercial systems have fixed specimen to target distances, and coating thicknesses are usually given by the manufacturer in nm/minute.

Argon is generally chosen as the plasma gas because it is inert, and produces ions which are much more massive than for example, nitrogen. This means more metal is dislodged by each ion impact and that, for a given system, the deposition time can be reduced, minimising specimen artefacts.

2.3.5. Coating Artefacts

Both vacuum evaporation and sputter coating can cause damage to delicate specimens. These artefacts are of three main types, thermal damage, surface etching and 'decoration' of the specimen.

Thermal damage in the evaporative technique is caused because of radiant heat from the tungsten wire during evaporation. It can be greatly reduced by keeping the heater/specimen distance as large as possible (>10 cm) and keeping the metal source as small as possible. In sputter coating there is little radiant heat produced, but the electrons produced, bombard the specimens, producing heat, and also etch the surface slightly. During sputtering these effects can raise the temperature of the specimens by as much as 40 °C, enough to cause pitting, melting and in extreme cases destruction of the specimens. Fortunately these effects can be greatly reduced by mounting the specimens on a cooled platform, and by fitting a small permanent magnet in the centre of the target, to deflect the electrons away from the specimens. Instruments with these modifications are sometimes called 'cool sputtering' units.

Surface etching of the specimens is more of a problem in sputtering than evaporation and is caused either by stray ions bombarding the specimen and sputtering material from it, or because of erosion by the impact of metal particles.

With microscopes capable of resolving finer and finer detail, SEMs are now revealing that the conventional coating techniques described so far can produce coatings made up of distinguishable particles or grains. These agglomerations or clumps of deposited material are often called 'decoration' artefacts. Recently new variations on sputter coating have been developed to produce much finer particles and less decoration artefacts. These systems use collimated beams of ions rather than a plasma cloud, to sputter materials.

2.3.6. Chemical Methods

Instead of improving the external surface conductivity of the specimens, several methods have been developed for improving their bulk conductivity. These chemical techniques have been applied, mostly to biological specimens. Basically it involves impregnating the samples with metal salts (usually osmium and/or manganese), often using organic metal ligands or mordants (eg thiocarbohydrazide). The metal complexes formed inside the specimens significantly increase conductivity. These methods have been extensively reviewed by Murphey (1978,1980).

SAQ 2.3a	Explain the principles of vacuum evaporation and sputter coating techniques. List their relative advantages and disadvantages.

SAQ 2.3a

SAQ 2.3b | Explain what is meant by 'coating artefact'. Explain how they arise and how they may be avoided or minimised.

2.4. CRYO-SEM

Over the last few years there have been several important developments in scanning electron microscopy. Probably the most important is that of 'cryo SEM', and this deserves mention.

The vacuum requirements of the SEM have posed one of the major limitations to the type of specimen which can be examined. Water, as we have seen, is a serious problem. However, if the hydrated samples are frozen very rapidly, to very low temperatures (< -190 °C), the water, now in its solid phase, poses fewer problems.

The formation of large ice crystals in specimens, particularly biological ones, can be very disruptive. Remember, slow cooling rates cause these large ice crystals to be formed, so several ingenious techniques, for very fast freezing are developing. Some rely on spraying the specimen with propane (cooled to -190 °C with liquid nitrogen), another involves 'slamming' the specimen at high speed into copper blocks cooled with liquid helium at -269 °C.

Once frozen the sample must be protected from condensation, and transferred to the cooled cryo-stage of the SEM. Providing the temperature can be kept well below -100 °C, the water has a low vapour pressure ($<10^{-3}$ Pa), will not evaporate (sublime) appreciably, and will not upset the vacuum.

There are several other advantages. For example the specimens are less likely to suffer thermal damage, when hit by energetic beams, so mass loss is reduced. In fact one of the major practical problems at the moment is one of mass gain. Precautions have to be taken to prevent the specimen acting as a condenser or cold finger. Most vacuum systems contain a few residual hydrocarbons from seals, pumps etc, and these will collect on the cold specimen, resulting in contamination, and mass gain.

This technique should open-up new areas of research, particularly for biologists and the food industry. For more details see Echlin *et al* (1978).

Open Learning 53

Summary

Generally externally applied coatings give the most reliable and consistent results, for routine SEM observation. Although not as good for observation work, evaporated carbon is particularly useful for samples for microanalysis. Chemical precipitation techniques are probably best used in combination with one of the other techniques, for difficult biological specimens, and are not at all suitable for microanalysis.

Vacuum evaporation is ideal for very thin metal coatings, and is the only way, at present, of depositing carbon. Sputter coating techniques have the important advantage that surfaces, hidden from the target, are coated almost as effectively as those which are line-of-sight. This means that there is no need to provide any sort of mechanical movement of the specimens in order to ensure even coating.

Objectives

You should now be able to:

- explain how samples are prepared for SEM examination;

- discuss the the principles and relative advantages of freeze-drying and critical point drying;

- explain why it is usually necessary to coat specimens and describe the principles of vacuum evaporation and sputter coating;

- list the various coating artefacts and explain their origins.

3. SEM X-ray Microanalysis – Instrumentation

3.1. INTRODUCTION

X-rays are produced whenever an electron beam interacts with matter, as in the SEM for example, and these can be used very effectively to provide us with information about the chemical composition of the specimens we are examining. Because basic X-ray microanalysers are self-contained, any SEM can be fitted with a system, usually with little or no modification. The detecting spectrometer is simply bolted to one of the spare vacuum 'ports' (holes) provided on most SEM specimen chambers. X-rays are collected and processed by independent instrumentation.

There are numerous techniques available for determining the elemental composition of samples, so what's different about X-ray microanalysis? Well, let's summarise its good points.

It can generally be regarded as a quantitative, non-destructive technique, that allows us to detect most elements, *in situ*, sometimes at levels as low as 10^{-19} g.

Its main limitations are that it cannot distinguish between ionic, non-ionic and isotopic species, and cannot at present detect the very low atomic number elements ($Z < 4$). It is a surface analytical technique, and because of the vacuum requirements of the SEM, is not very suitable for hydrated specimens. We'll look at the system's limitations in more detail (3.5) when we have a better understanding of how it works.

3.2. X-RAY PRODUCTION IN THE SEM

We learnt earlier (1.3) that, among other effects, X-rays are produced when the electron beam bombards the specimen. Now we will look at this effect in detail, and find out how it can be used as a powerful analytical technique.

In order to understand X-ray production we need to look at the arrangement of electrons around the atomic nucleus, ie the Bohr model of the atom (this ought to be familiar to you, and is intended only as revision). As an example let's use a magnesium atom. Fig. 3.2a represents the magnesium atom in its normal 'ground' state.

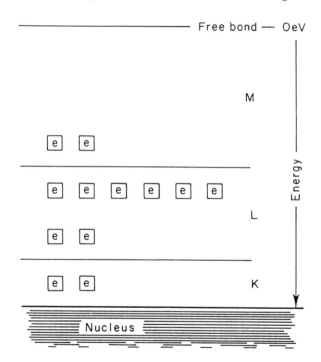

Fig. 3.2a. *Magnesium atom in ground state*

The orbiting electrons are arranged in distinct energy bands or shells. The 'principal quantum number' (n), is used to indicate a shell in which all the electrons have nearly the same energy ($n = 1$

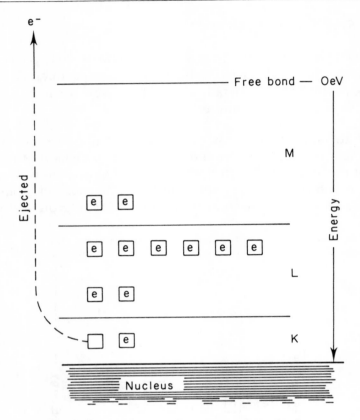

Fig. 3.2b. *Magnesium atom in exited state*

being nearest to the nucleus and so on). In X-ray work these shells are referred to as K, L, M etc. Nearest the nucleus, corresponding to $n = 1$, is the K electron shell, above it, further from the nucleus, and therefore at lower energy levels, are the L, M, N shells (corresponding to $n = 2,3,4$ etc). The electrons in the innermost (K) shell require the most energy to remove them from the atom. Remembering that the atomic number of magnesium is 12, this means that the K shell is filled with two electrons and the L shell is filled with eight. The M shell however, has only two electrons present, leaving sixteen vacant.

A beam electron, accelerated towards the atom, may interact with

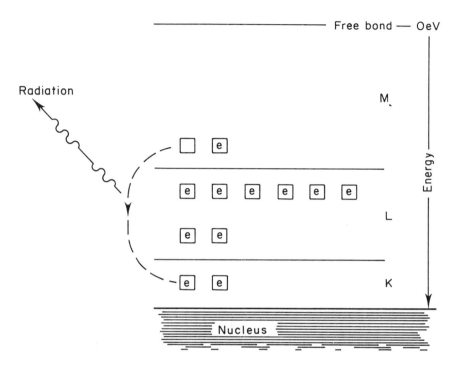

Fig. 3.2c. *Magnesium atom in relaxation state*

one of the shell electrons, causing it to be ejected. Our atom is left in an ionised or 'exited' state (Fig. 3.2b).

Suppose the beam electron had sufficient energy to knock out one of the K shell electrons. Because the vacancy created is at a higher energy, an electron from an outer shell, in this case L or M, will fall into the lower level vacancy, restoring the atom to the 'ground' (non-exited) state (Fig. 3.2c). This 'relaxation' stage releases an amount of energy equal to the difference in energy between the shells involved in the transistion.

One possibility, and the one that is of interest here, is that the energy is released as a photon of electromagnetic radiation. If the ejected electron was from an inner shell (K,L,M) then the energy is such that the photon is an X-ray. These discrete packages of energy are called quanta (singular quantum).

From our diagram it is obvious that there are several ways in which this relaxation can occur. An M shell electron may fall directly down to fill the K shell vacancy, or an L shell electron may do so, creating an L shell vacancy, which in turn may be filled by an M shell electron.

3.2.2. Characteristic X-rays

Although shells have distinct energies, within these shells the energy of the electrons varies with the atomic number, because the positive charge on the nucleus is also dependent on the atomic number. Even between elements of adjacent atomic number this variation in energy is significant. From this we can see that whenever an electron transition occurs the energy released will be unique for each atomic number, and therefore, unique to each element.

Assuming the energy quanta are X-rays, they will have distinct energies which can be expressed in electron-volts (eV), or thousands of electron-volts (keV). Just as visible light photons have a wavelength (λ), so too do the high energy X-ray photons, and this is related to their energy:

$$\text{wavelength } \lambda = hc/E \text{ nm}$$

where h is the Planck constant, c the speed of light, E the energy in (keV), and the wavelength is given in nanometres (nm).

SAQ 3.2a Using the values of the constants, h and c determine the numerical relationship between the wavelength, λ (expressed in nm) and the energy E (expressed in J and in kev) in the equation

$$\lambda = hc/E$$

$h = 6.626 \times 10^{-34}$ J s; $c = 2.998 \times 10^8$ m s^{-1}; 1 eV $\equiv 1.602 \times 10^{-19}$ J

SAQ 3.2a

Fig. 3.2d shows the electromagnetic spectrum and the relationship between wavelength, quantum energy and frequency. It will be useful to remember that high energy X-rays have shorter wavelengths than low energy ones.

An instrument capable of analysing the physical characteristics, either wavelength or energy, of these X-rays, is capable of identifying the element which produced them.

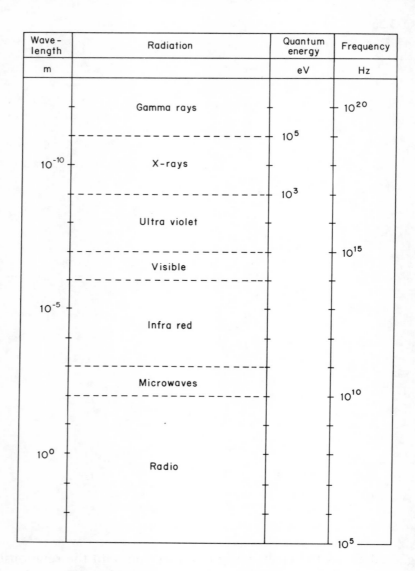

Fig. 3.2d. *The electromagnetic spectrum*

SAQ 3.2b | For an X-ray of wavelength 2.5×10^{-10} m, calculate the quantum energy E (in keV) and the frequency of the radiation.

3.2.3. X-ray Continuum

We mentioned earlier (1.3) that there is another possible result of specimen/beam interaction – X-ray continuum emission. Incident beam electrons can be slowed down as they pass through the electrostatic (coulombic) field, close to the atomic nuclei and inner electron shells. In doing so they give up some of their energy to produce X-rays. Some beam electrons may be unaffected, whilst others are slowed dramatically, or even stopped, so the energy (and therefore the wavelengths) of the resulting X-rays is infinitely variable between zero and the maximum beam accelerating voltage. This wide band of X-ray 'noise' is usually called the X-ray continuum or bremsstrahlung ('braking radiation').

If we refer back to Fig. 3.2d for a moment, we can see that if for example a 20 keV incident electron was slowed only slightly then the energy loss is only a few eV, so the radiation would not be an X-ray at all, it would be in the visible part of the spectrum. On the other hand, the same 20 keV electron slowed almost to rest would give up most of its energy, producing say a 19 keV X-ray. In practice most of these energy losses are below 1 keV. As far as practical microanalysis is concerned the continuum appears undesirable since it represents a background signal, on which the useful, characteristic X-rays, are superimposed. The continuum does contain information about the mean atomic number of specimens and can play a useful part in correction routines in some quantitative systems.

3.2.4. Families of X-ray Lines

Earlier in this Part, when considering the magnesium atom, we discovered that when a vacancy is created (eg in the K shell), electrons may fill it directly (in this example, from the M shell), or intermediate transitions may occur (L to K and M to L). In the simple case, where one transition occurs, a single characteristic X-ray peak or 'line' is produced. Where two transitions occur two distinctly different X-ray lines are produced. The lighter elements have only a few possible transitions, but the heavy elements, with their very complex shell structures have numerous possibilities.

It is important for us to be able to sort out distinct groups or 'families' of X-ray lines, if we are to analyse them!

Ionisation can be caused by creating vacancies in any of the inner shells (K, L, M etc), so the X-rays produced occupy distinct energy (and wavelength) bands, just as the electrons occupy bands in the shells. These bands are used to divide the X-ray lines into major families. For example, the X-rays produced by electrons filling a vacancy in the K shell are designated K lines, those produced by electrons filling a vacancy in the L shell are designated L lines and so on.

These families of lines are further subdivided according to which shell provided the electron to fill the vacancy. In this example an

L shell electron filling the vacancy would produce a Kα X-ray, and an M shell electron would produce a Kβ.

∏ Which X-ray lines would you expect to be produced from:

(i) carbon (atomic number 6)
(ii) magnesium (atomic number 12)

Because carbon (atomic number 6) has only two K and four L shell electrons, the *only* possible transistion is from L to K, so the *only* possible X-ray is the Kα. Although it is possible for an L shell electron to be ejected by collision, there are no M shell electrons to fill the vacancy.

Magnesium (atomic number 12), on the other hand, has two K, eight L and two M shell electrons. In this case a vacancy created in the K shell could give rise to either Kα *or* Kβ. In addition, a

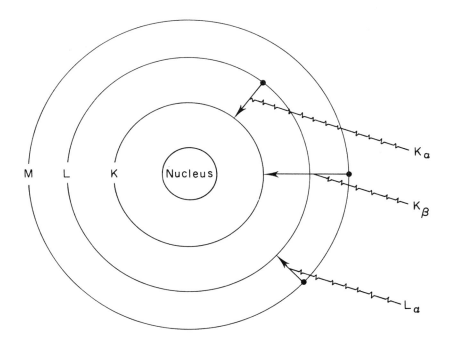

Fig. 3.2e. *Some of the possible electron transitions*

vacancy created in the L shell, could now be filled from the M shell, producing an Lα X-ray. Fig. 3.2e shows these possibilities.

If you find this confusing, try the exercise with lead!

Although many publications subdivide X-ray families still further, it is unnecessary for us to do so.

3.3. WAVELENGTH DISPERSIVE SYSTEMS (WDS)

Sometimes referred to as WDS (wavelength dispersive spectrometry) or WDX (wavelength dispersive X-ray) analysis, this was the earliest type of SEM microanalysis. As the name implies this type of instrument determines the specimen's elemental composition, by analysing the wavelength of the X-rays emitted from it.

First we will look at how the X-rays are sorted (in terms of their wavelength) and then how they are detected. We will then discuss how this information can be used to determine composition. An overall diagram of the system is shown in Fig. 3.3a.

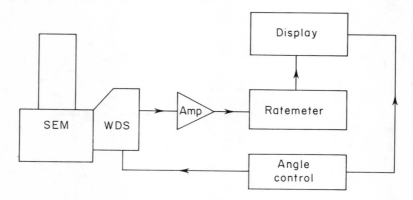

Fig. 3.3a.   *Overall WDS system*

3.3.1. Analysing Crystal

Separating the X-ray spectrum relies on diffraction. Some of the X-rays generated from the SEM specimen, leave it at an angle ϕ, the 'take-off' angle, pass through collimating slits, and are allowed to strike a crystal, which has its atoms arranged in very ordered planes. In Fig. 3.3c the crystal has lattice spacings (d), and the X-rays are incident on the crystal at an angle θ. Bragg's law dictates that only X-rays with a wavelength λ will emerge from the crystal, where:

$$n/\lambda = 2d \sin \theta$$

where, n is an integer. All other wavelengths will either be absorbed by the crystal or pass through it and be lost.

See Goodhew (1974), or any textbook on crystallography for a detailed explanation of diffraction and Bragg's law. Effectively the crystal behaves rather like an X-ray 'mirror', where changing the angle of incidence, θ, changes the wavelength of the X-rays 'reflected'. Now we are able to select which of the characteristic X-rays we are going to quantify.

To detect the X-rays we use a gas flow proportional detector, which converts them into an electrical signal.

3.3.2. Gas Flow Proportional Detectors

These detectors are close relatives of the Geiger tube, familiar to most students. Shown in Fig. 3.3b, it consists of a gas-filled tube (T), containing a thin wire electrode (E). Usually made of tungsten, this wire is insulated from the tube and held at about 3 keV potential. There is a very thin window (W), through which X-ray photons can pass into the tube. When an X-ray photon passes through the gas in the tube it is usually absorbed by the gas, releasing a photo-electron. This quickly loses its energy by ionising other gas atoms, releasing yet more electrons. All these free electrons are attracted to the positive electrode wire, where they produce a current pulse. Providing the applied potential is carefully controlled, the device can be made to operate as a 'proportional' detector, ie the output current pulse is proportional to the energy of the incident X-ray photon. Generally there are two small pipes (P) connected to the main tube of the detector, to provide a flow of gas over the electrode. A constant supply of gas is fed through the system so that the ionization can occur. A mixture of 90% argon and 10% methane is commonly used in these so called constant flow proportional detectors.

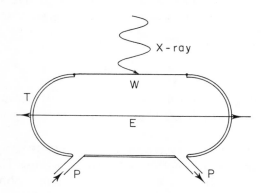

Fig. 3.3b. *Gas flow proportional detector*

A constant flow of gas is a necessary complication because it is difficult to produce very thin windows which can withstand the pressure difference (about 1 atm) without leaking, and failure. Keeping the detector in the SEM's vacuum is necessary to reduce absorption of X-rays by air, and a pressure of at least 1 atm is required for reasonable detector sensitivity.

Just to illustrate the point, let's look for a moment at the construction of the window (W). Beryllium foil is commonly used and, even if it were only 34 μm thick it would absorb 98.8% of the X-rays from, for example, aluminium (Hendee *et al* (1956)). Reducing the thickness to 7.5 μm would still absorb 45%. Plastic materials such as Mylar and Formvar offer improved transmission, at thicknesses of around 1–2 μm, but these need supporting over wire meshes, which tends to reduce their efficiency. It is quite amazing that films of this thickness can withstand such pressure differences at all!

3.3.3. Overall System

Let's look at how a complete, practical instrument works. Precision engineering is required to enable us to alter and accurately measure the angle of the diffracting crystal. Knowing this, the Bragg angle allows us to calculate the wavelength of the emergent X-ray.

WD spectrometers are bulky and their geometry is quite complex. Crystals and detectors are moved through focusing 'circles' rather than by simple tilting, and the crystal is shaped in a curved rather than flat, planar form.

From Fig. 3.2d (in the previous section), we can see that the spectrum of interest to us extends from about 0.1–1 nm. A single diffracting crystal would not be able to cover this range, because of the practical engineering problems involved in rotating it through large angles. To overcome this difficulty, most spectrometers are fitted with several crystals of different composition and lattice spacing, so that the entire range can be covered. Crystals of lithium fluoride,

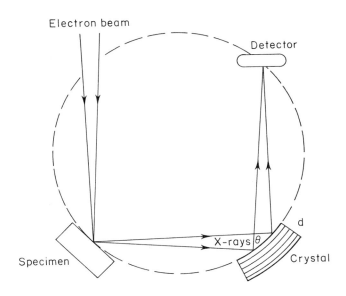

Fig. 3.3c. *Wavelength dispersive spectrometer*

rubidium acid phthalate and α-quartz are among the compounds commonly used.

In a practical instrument (Fig. 3.3a), fixing the angle of the crystal (of known lattice spacing), allows us to determine which X-rays and therefore which elements we detect.

Once an X-ray has triggered the detector, the output signal is amplified and passed to a pulse shaping circuit, where its size and shape are standardised, ready for counting. In early instruments these shaped pulses were applied to a ratemeter, and displayed on a chart recorder, making interpretation awkward.

Modern instruments use digital computers to count the pulses, and apply sophisticated statistics and corrections to the results (Section 3.6). To summarise then, wavelength and elemental composition can be calculated from the crystal angle and lattice spacing, and elemental concentration from the X-ray count rate.

One final point, each spectrometer is only capable of analysing one element at a time, because the diffracting crystal is fixed at a particular angle for that element, and although this might seem trivial, it does impose limitations.

3.4. ENERGY DISPERSIVE X-RAY MICROANALYSIS (EDX)

We've seen how wavelength can be used to determine composition, EDX microanalysis offers an alternative approach.

Currently the most commonly used of the two systems, EDX has developed from the work of Fitzgerald, Keil and Heinrich, who, in 1968 published details of a system based on the use of a 'solid-state' detector.

3.4.1. Solid State X-ray Detector

These devices have much in common with the solid state electron detector we looked at earlier (1.4.1) and Fig. 3.4a shows its construction. The semi-conducting crystal (C) is the key component. A typical detector crystal has an active area of about 4 mm diameter and is about 3 mm thick, although the exact geometry is quite complex and very precise.

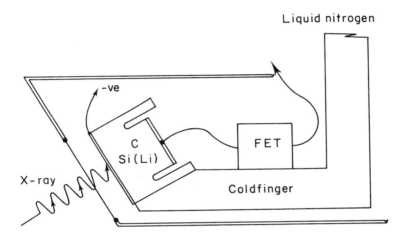

Fig. 3.4a. *Solid state X-ray detector*

As with the electron detector, silicon is again used, and in this case is doped with impurities of lithium and boron. The manufacturing process involves very controlled diffusion of minute quantities of lithium into one surface of a single crystal of p-type silicon. The process is known as 'drifting', and as a result this type of detector is often called a 'Si(Li) drifted' detector. Very thin (20 nm) gold contacts are then evaporated onto each end of the crystal.

Very simply, the result is a reversed bias n–p diode, but unlike many semiconductor devices it has to be able to withstand very high electric fields (>1000 V cm^{-1}) for it to operate.

Unfortunately, if this high voltage were applied at room temperature, the lithium would become very mobile, tending to redistribute itself within the silicon and, as a result, the detector performance would be permanently impaired. Although this means we have to cool the crystal (usually to about −190 °C), we shall see later that this is essential for other reasons.

Now let's look at how it works. If an X-ray strikes the front face of the crystal it gives up its energy by collision events. A single photon colliding with the crystal will be absorbed by a silicon atom, releasing an electron. This energetic free electron scatters inelastically through the silicon producing free electrons and 'holes' (electron vacancies in the valance band). The excited silicon atom may release its excess energy by producing an Auger electron, which in turn produces more electron/hole pairs. Alternatively, it may do so by releasing a (silicon) X-ray, which itself restarts events. Multiple events like these occur inside the crystal, until all the energy is absorbed within it. With a field in excess of 1 keV applied to the gold contacts the electron/hole pairs are separated, each migrating in opposite directions, and a current pulse is formed. Although numerous, the events triggered by a single photon, produce a pulse lasting less than 1 μs. It is worth mentioning that energetic (eg backscattered) electrons also trigger the detector.

Semiconductor physics dictates that on average 3.8 eV of energy is required to create one electron/hole pair in silicon. We can easily calculate the number of electron/hole pairs (n), and hence the current induced in the detector, using:

$$n = E/e$$

where E is the energy of the incident X-ray, and e is a constant (3.8 eV) for silicon. From this it is obvious that the charge pulse from the detector is directly proportional to the energy of the incident X-ray photon.

∏ Calculate the number of electron/hole pairs resulting from a single 5 keV X-ray photon striking the detector. Assuming the pulse produced lasted 10 microseconds calculate the resulting current.

$n = 5000/3.8 = 1300$ electron/hole pairs

Open Learning

This represents a charge of only 2×10^{-16} C,

ie electronic charge $\times n = 1.602 \times 10^{-19} \times 1300$

$$= 2 \times 10^{-16} C$$

Assuming the current pulse lasts 10 microseconds the current is given by:

$$2 \times 10^{-16}/10^{-5} = 2 \times 10^{-11} A$$

$$= 20 \text{ pA}$$

A very minute current indeed! To make such a small signal usable it must be amplified by 10^{10} or more.

First, the current is applied to a specially designed field-effect transistor (FET), mounted very close to the detector. This is the detector pre-amplifier. As well as the signal electrons, random electron 'noise' in the crystal and electronics would also be amplified and, unless steps are taken to limit it, might overwhelm the signal. Immersing the crystal and pre-amplifier in liquid nitrogen in a cryostat, cools the system (to about -190 °C) and greatly improves the signal/noise ratio. Remember, this was a requirement anyway, to prevent lithium migration in the silicon. Ideally the detecting crystal should be exposed directly to the source of X-rays, ie it should be adjacent to the SEM specimen, but in practice this presents problems. The detector may become contaminated by hydrocarbons etc (from the microscope vacuum system) condensing on its cold surfaces. Such 'windowless' detectors are available where very high performance is essential. Far more common however, are the 'window' types, in which the end of the detector tube is sealed with a window of thin (8 μm) beryllium foil. This does absorb some X-rays, but makes a much more rugged detector, protected from contamination and sealed in its own ultra high vacuum.

One further advantage is that the beryllium window blocks any backscattered or secondary electrons from entering the detector and causing 'noise'.

Having passed through the pre-amp stage the signal is large enough to leave the cryostat, and pass to the main amplifier and processor electronics.

3.4.2. Processing Electronics

Each pulse sent to the FET pre-amplifier, is added to the previous pulse, so as to produce a 'staircase' waveform. Each step in the waveform has a height proportional to charge and hence X-ray energy. To ensure precise proportionality the waveform is restored to zero when a preset height is reached. For those with a keen interest in electronics, circuits of this type are almost immune to the 'noise' usually found in resistive feedback circuits.

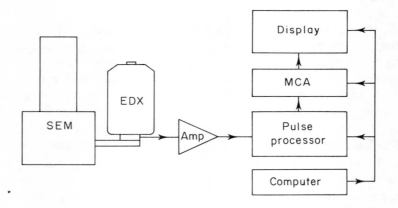

Fig. 3.4b.   *Overall EDX System*

The main amplifier/pulse processor converts the analogue staircase waveform to a string of digital pulses of identical width. Although amplified, the relative amplitude (height) of the individual pulses remains unchanged, and so still contains the information about the X-ray energies.

The string of pulses now passes to a multi-channel analyser (MCA). We can regard this as a 'black-box' which separates the pulses according to their amplitude, counts them and stores the results. It

Open Learning 73

rather like an electronic sorting office – each pulse is identified according to its height, put into a numbered box ('channel'), and the box number is stored together with the number of pulses in that particular box.

Practical analysers currently have about a thousand channels. This means that if, for example, the SEM's accelerating voltage is 20 keV, the X-ray spectrum spans 20 keV, so each one of the thousand channels in the MCA is separated by 20 eV. Modern computing systems are capable of providing more channels, but this would not improve the instrument's resolution, as we shall see later (3.5), it is the detector that is limiting.

Once collected, sorted and stored all the data can be manipulated quickly and easily by a built-in computer, and the results displayed in a variety of ways (Section 3.6).

SAQ 3.4a Explain the principles underlying the detection of X-rays and list the advantages and disadvantages of WDS and EDX systems.

3.5. OPERATING CONDITIONS AND LIMITATIONS

Probably the first question asked by potential users is – what are the limits of detectability? Certainly this is the most difficult one to answer, because it is dependent on a number of things, specimen, operating conditions etc. Fig. 3.5a is intended only as a guide to the detection limits for the two principal X-ray analytical techniques. The figures are can really only be taken as the 'best possible' rather than 'always attainable'.

	WDS	EDX
Elements detectable	$Z > 4$	$Z > 11$ (Be Window detector)
		$Z > 5$ (windowless detector)
Minimum conc. (ppm)	220*–750	1000
Minimum sample volume	10^8 nm^3	
Minimum mass	10^{-18} g	
Spatial resolution		> 25 nm
Sampling depth		< 500 nm
Relative quantitative accuracy ($Z > 11$)		1–10% (using (zaf†)

* This lower value is based on work carried out by Rick, Dorge and Thurau, 1982.

† ZAF is a quantitative correction technique and is discussed later (Section 3.6).

Fig. 3.5a. *Comparison of WDS and EDS detectability limits*

If we examine some of the problems encountered during microanalysis we may begin to understand why it is difficult to be emphatic about detectability limits.

3.5.1. Mass Loss

One of the problems with SEM analysis, particularly on biological material is that the specimen will lose mass when bombarded by electrons. Reducing the beam current to very low levels does not completely overcome the problem. The mechanics are not well understood, but calculations show, that with uncooled biological specimens, 20–40% losses are not uncommon (Hall and Gupta 1974). Such unpredictable variations pose serious problems. Cryo-SEM where specimens can be maintained below -100 °C, seems the most promising solution at present.

3.5.2. Spectrometer Resolution

Resolution in X-ray spectrometers is a measure of their ability to separate adjacent X-ray lines. High resolution is very desirable since there are fewer problems associated with overlapping peaks (Section 3.6). WDS offers resolutions of around 10 eV, and is very much better than that currently available by EDX, where it is energy dependent, and is typically only about 150 eV (at 5.9 keV).

3.5.3. Minimum Probe Size

This is the smallest electron beam diameter which can be used for practical analysis. Remember, that although dependent on probe diameter, specimen/electron interactions occur from a much larger interaction volume (Section 1.3). WDS spectrometers are relatively inefficient X-ray collectors, so to produce a beam with sufficient energy to excite X-rays, a minimum diameter of about 200 nm is required. Conversely, the more efficient EDX detectors, mean that lower beam currents can be used, so the beam diameter can be reduced to a minimum of about 5 nm (SEM permitting). Reduced beam diameters and currents also help minimise mass loss, and is particularly important for light element specimens.

Very low beam currents and small beam diameters yield very low X-ray counts, and collection times may have to be increased dramatically, to obtain meaningful results.

3.5.4. Escape Peaks

These are exclusive to EDX detectors, and can be misleading. X-rays with energies >1.84 keV may excite X-rays in the silicon of the detector. Some of these silicon Kα X-rays (energy 1.74 keV) may escape through the window. This produces a spurious 'escape peak' which appears exactly 1.74 keV below a genuine major element peak, adding further complications.

3.6. DATA HANDLING

So far we have only collected data from the spectrometers consisting of either a wavelength or energy and a count-rate. We also know that count-rate is proportional to concentration, providing certain operating conditions are kept constant.

We must now look at ways of using this information to greatest effect, and presenting it in a convenient form. There are two possible approaches, either qualitatively or quantitatively.

3.6.1. Qualitative Analysis

There are many occasions where simple qualitative analysis can provide useful information. When dealing with microscopic particles too small for normal chemical analysis, it is often only necessary to establish the presence or absence of a particular element.

In this case all that is required is a complete X-ray spectrum. Plotting count-rate against wavelength or energy would produce a print-out similar to that shown in Fig. 3.6a. Tables of energies and/or wavelengths (Johnson and White, 1970) can be used to identify the characteristic X-ray peaks, although modern computer based systems have a data base from which they can identify and label peaks automatically.

Open Learning 77

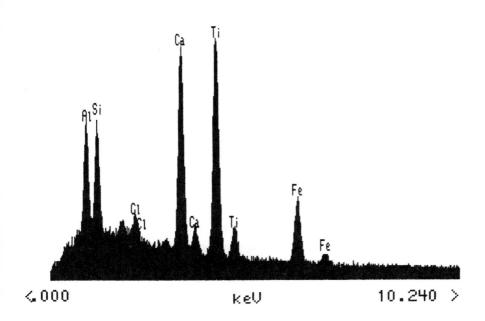

Fig. 3.6a. *Typical X-ray spectrum (EDX)*

It is important to realise that only the area of the specimen being scanned by the electron beam will be analysed. If a normal image is being acquired, then X-rays will be collected from the entire field of view. The actual area will of course depend on the magnification (Section 1.2). There are occasions when it is useful to analyse, for a specific element, along a line or transect across the specimen. Such 'linescans' are easily achieved by switching off the vertical frame (Y) scan generator on the SEM. Now the electron beam will repeatedly scan over the same, single, horizontal line, generating X-rays as it goes. Plotting count rate along the line shows the concentration of the element at each point along it.

Carrying this idea one step further allows us to do point or 'spot' analyses. Switching off both the vertical frame and horizontal line (X) scan generators, leaves the electron beam stationary, generating X-rays and analysing from a single point on the specimen.

Π   Can you suggest what determines the sampling size on the specimen?

Spot size (electron beam diameter) determines the area of the surface irradiated, but the 'interaction volume' also depends upon other factors, eg accelerating voltage, specimen thickness, mean atomic mass and mean atomic number.

Sometimes distribution of elements is much more important than concentration, and this has led to an X-ray mapping technique. In this sort of application, large volumes of data are being handled quickly, and it is not possible at the moment to apply sophisticated data correction routines to quantify the results.

Let's look at the 'mapping' technique, and see how it can be used. Suppose we have a specimen which is not homogeneous, eg a platinum electrode which has different metals electro-deposited on its surface. Determining the distribution of the metals is simple for a modern EDX or WDS system.

Using just the SEM, we would certainly be able to image the surface and might be able to distinguish some compositional difference (atomic number contrast), if we used a backscattered detector. With an X-ray analyser we can detect which elements are present, where they are located on the specimen, and get some idea of relative concentration. If the elements present in the specimen are unknown to us, we first carry out a basic qualitative analysis, ie collect X-rays over the entire range of energies or wavelengths. From this we can choose which of the elements detected will be used in our map. The analyser is then set-up to detect these elements. In a WDS system the spectrometer is set to the appropriate wavelength (ie crystal and angle). In the EDX system the channels covering the required energy range are selected (this is often referred to as setting a 'window'). Now the mapping can begin. The SEM is made to scan its electron beam slowly over the specimen, in the usual way. But, instead of using the signal from an electron detector to form the image on the viewing screen, we use the output from the X-ray spectrometer. As a result the only points made bright on the display, will be those where our specified element is located. The resulting 'dot-map' consists of bright points showing the location of the particular element, and

(i) Lead-Free Paint Sample ## (ii) Lead-Based Paint Sample

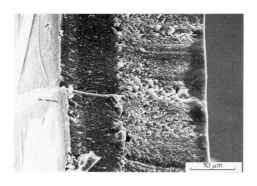 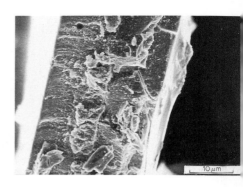

Secondary Electron Image　　　*Secondary Electron Image*

 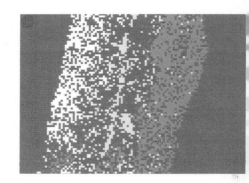

X-Ray Elemental Map　　　*X-Ray Elemental Map*

Yellow = Calcium　　Blue = Aluminium　　Green = Iron

White = Titanium　　Red = Lead

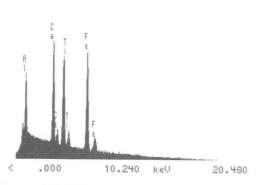 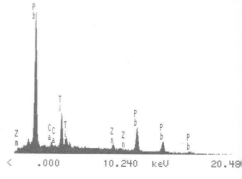

EDX Spectrum　　　*EDX Spectrum*

Fig. 3.6b. *Quantitative X-ray spectra and elemental maps for flakes of dried paint, (i) lead-free paint, (ii) lead-based paint*

the brightness of the points gives an indication of its concentration. Modern systems allow dot-maps from different elements or combinations of elements, to be colour-coded, and superimposed on the normal SEM image. The result is a detailed surface image showing chemical composition and distribution.

Fig. 3.6b shows the quantitative X-ray spectra together with elemental maps obtained from two microscopic (<1 mm^3) flakes of dried paint. Although both images appear similar the spectra are quite different. The maps show each flake to have three distinct layers. That for the lead-free sample shows that an aluminium primer, calcium undercoat and iron based top-coat were used. On the other hand the map for the lead-based paint shows the presence of a lead-based primer, calcium undercoat and titanium top-coat.

The distribution of platinum in a plant root tip as determined by WDX and EDX is shown in Fig. 3.6c.

3.6.2. Quantitative Analysis

Unfortunately, we cannot use the X-ray count-rate as a direct indication of concentration. The physics of X-ray emission is very complex, much more information about the system's geometry and about the behaviour of the specimen and its constituent elements is needed. Different X-ray lines from the same element have different relative intensities (peak heights), which must also be taken into account. More problems arise because it is often not possible to use calibration standards identical to the specimen composition. Complex mathematics and statistics must therefore be applied, to transform the raw data into fully quantitative results. It is beyond the scope of this text to deal with the mathematics involved, particularly since there are several excellent publications which deal with the subject in depth (Goldstein *et al*, 1981). Besides with modern computer-based systems, an awareness of the methods and limitations is really all that is required in order to achieve reliable results.

Fig. 3.6d shows the spectra and quantitative data obtained from three types of asbestos fibres. Quantitative chemical data is essential for positive classification (physical appearance and spectra shape

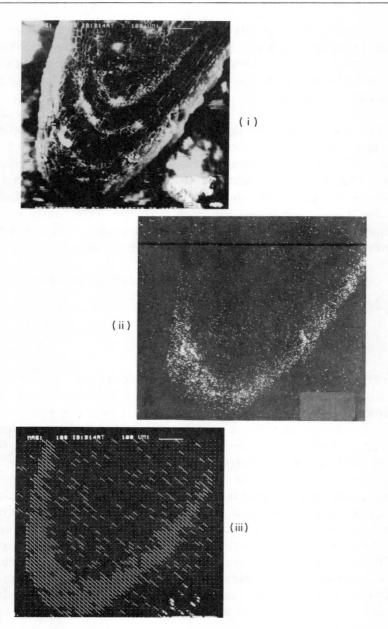

Fig. 3.6c. *Distribution of platinum in plant root tip. (i) SEM of root tip, (ii) platinum distribution using WDX and (iii) distribution using EDX. (P. Parsons, PhD. thesis, University of London)*

are not conclusive). Identifying and classifying asbestos fibres is an important application for X-ray microanalysis. A single fibre only a few micrometres long can be identified from, for example, lung tissue or environmental pollution monitoring equipment.

Previously (Section 3.5), we introduced the concept of specimen mass/thickness and topography in relation to X-ray emission. These play an important part in quantitative analysis. With thin specimens, X-rays can usually be regarded as having come from an area of the same diameter as the incident electron beam. Thick, bulk specimens are more difficult because electrons interact with a much larger volume of the specimen. This 'interaction' volume is dependent on several factors such as beam kV, specimen atomic number etc. Bulk samples would produce much larger errors than thin specimens in quantitative work, so some mathematical correction is essential for thick specimens. By far the most important technique for quantifying results is called ZAF.

ZAF – is an acronym from the three separate effects; atomic number (Z), absorption (A) and fluorescence (F), which the method compensates for.

We know that when an incident electron enters the sample it may undergo elastic scattering (Section 1.3), in which case it may re-emerge before it has excited any X-rays. Similarly, X-rays may not be generated because the electron may lose energy so rapidly, as it penetrates deeper into the surface, that it soon has insufficient energy. Both of these effects are dependant on the atomic number of the sample, and if left uncorrected could cause errors in excess of 10%.

An explanation of the absorption effect is quite straightforward. X-rays generated at the surface of the specimen are emitted without loss, whereas those, produced from deeper within the sample, have further to travel through the sample and are likely to be absorbed before they can escape to the detector. The take-off angle, between the specimen surface and detector must be known, since it is important in determining the path length of the X-ray through the specimen, and the probability of absorption (Fig. 3.6e). Absorption corrections are themselves complex, and must take into account

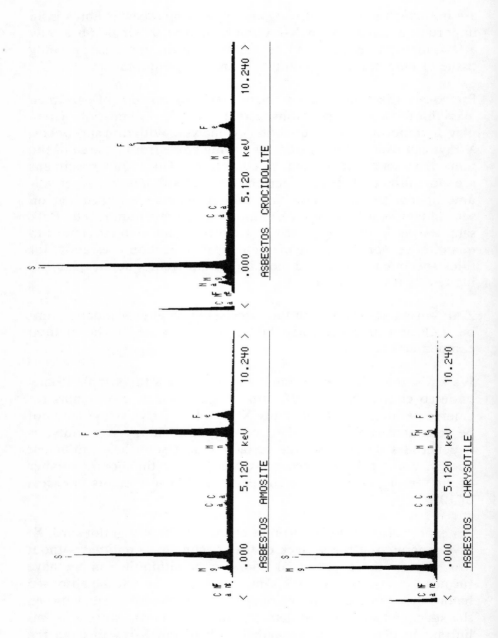

Fig. 3.6d. *Spectra and quantitative data for three types of asbestos fibre. (i) amosite, (ii) chrysotile, (iii) crocidolite*

	Weight % in:		
	Amosite	Chrysotile	Crocidolite
SiO_2	50.6	42.6	53.7
Na_2O	0.1	0.4	4.4
CaO	0.7	0	0.3
MgO	6.5	41.9	2.6
MnO	1.6	0	0
FeO	40.1	2.3	35.5

numerous factors eg the incident electron's kV, take-off angle, mean atomic number and mean atomic mass of the specimen etc. In very thin samples the effect is almost negligible.

X-rays produced within the specimen may interact with some of the inner shell orbital electrons, producing further 'secondary' X-rays. The effect causes abnormally high counts in the lower energies. Although generally considered the least important of the three corrections secondary fluorescence can cause errors as large as 15%, when analysing elements of adjacent atomic numbers.

(*a*) Peak deconvolution

Further errors result from the fact that X-ray peaks frequently overlap each other, the problem is noticeably worse for EDS than for WDS, because of the poorer resolution. As an example, when analysing for calcium, the major X-ray peak is the Kα. Fig. 3.6f (from Goldstein *et al*, 1981), shows some of the peak overlaps for a few of the more common elements, it is not comprehensive. If the sample contains potassium, the Ca Kα peak is exaggerated, because potassium produces a minor peak (Kβ), which overlaps it. Unscrambling adjacent overlapping peaks is known variously as 'peak-stripping', 'peak unravelling' or more commonly 'peak deconvolution'. Several methods have evolved for doing this, most require the use of standard calibration specimens, which do not contain overlapping elements. These are used by the computer to calculate

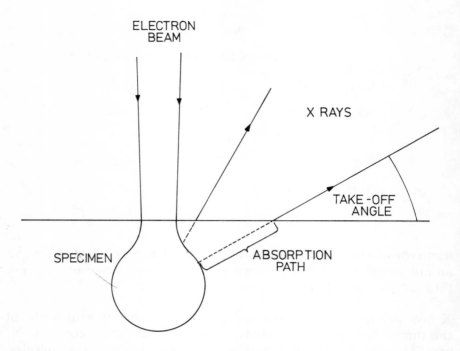

Fig. 3.6e. *Effect of take-off angle on absorption path length*

and construct ideal peaks, located precisely on the spectrum, and of perfect shape. The resulting library of elemental profiles can then be subtracted from the convoluted spectra to restore the data. This technique is known as 'fitting of least squares' or FLS.

(*b*) Background subtraction

Ideally X-ray peaks would be due exclusively to the specific elements in the sample, this is not the case. X-ray spectra do not have a zero base line, but a varying background count. This is caused by the X-ray continuum from the specimen (Section 3.2), inherent noise in the system and stray X-rays produced from the component parts of the SEM stage etc. Instead of having a flat base line we have a variable non-linear background. Fig. 3.6a shows that the background intensity is highest at the lower end of the spectrum (<5 keV). This means that the background is more significant when

interfering element	line	element affected	line
Ag	L	Cl	Kα
Ag	L	K	Kα
As	L	Na	Kα
As	L	Mg	Kα
As	L	Al	Kα
Ba	Lα	Ti	Kα
Cu	L	Na	Kα
Cr	Kβ	Mn	Kα
Fe	Kβ	Co	Kα
K	Kβ	Ca	Kα
Pb	M	S	Kα
Pb	M	Cl	Kα
Pb	M	Mo	Lα
Pb	L	As	Kα
Pb	L	Se	Kα
Zn	Lα	Na	Kα

Fig. 3.6f. *Interference between elements (Based on Goldstein et al, 1981)*

the lighter elements are being sought. Unless the background count is subtracted, peaks are exaggerated, and quantitatively inaccurate.

3.7. OTHER SURFACE ANALYTICAL TECHNIQUES

You should be aware that there are a number of other related techniques capable of providing information about the elemental composition of surfaces. It is beyond the scope of this text to cover these, but for completeness they are listed:

X-ray photoelectron spectroscopy (XPS) – sometimes known as electron spectroscopy for chemical analysis (ESCA);

Auger electron spectroscopy (AES) and scanning Auger microprobe (SAM);

ion scattering spectroscopy (ISS);

secondary ion mass spectroscopy (SIMS);

electron energy loss spectroscopy (EELS) – suitable only for very thin specimens.

SUMMARY AND OBJECTIVES

Summary

We have seen that most of the major correction factors mentioned above are inter-related, and where atomic number is involved in the calculations, seem to suggest that we need to know what elements are present in the sample before we can begin. This is partially true, and the highly developed systems in use today are 'iterative' systems. This means that they carry out an initial analysis, to determine which elements are present. Corrections are then applied to the data, taking into account which elements are present. This data is then further refined, taking into account the corrections for element concentration, then corrected again, and so on, until little or no improvement results from further processing.

By now it will be apparent that extracting quantitative data from EDX and WDS microanalysers is rather involved. Nowadays, as far as the operator is concerned, it is usually just a case of supplying the system's computer with the appropriate data (eg tilt angle, keV etc) when prompted. However, considerable experience may be required in assessing the reliability and accuracy of the resulting quantitative data.

Continuing research is providing a better understanding of the principles and limitations of X-ray microanalysis, and very rapid advances are being made in quantitation. More accurate and efficient computer software, combined with cheaper, faster and more powerful computers will ensure that the quantitative X-ray microanalyser will, like the SEM, become a routine tool in many areas of science and industry.

Objectives

You should now be able to:

- explain the processes involved in X-ray production;

- describe the two principal types of detection system, and the practical ways in which they can be used to provide elemental composition;

- discuss the problems associated with quantitative microanalysis;

- outline the limitations of X-ray microanalysis.

Self Assessment Questions and Responses

SAQ 1.2a — It is quite common to use the SEM at a magnification of 40 000×, so assuming that we retain the 20 cm square CRT screen, work out the area being scanned in the column at this magnification.

Response

At 40 000 times, the area scanned in the column is given by:

screen width/magnification = 0.2/40 000 m

= 5 μm

Thus the area scanned is 5 μm × 5μm.

If you got this wrong, you probably muddled the units of length. Some people find it easier to convert all measurements to metres first. It is worth remembering that we are using linear magnifications, not areas.

****

Open Learning 89

SAQ 1.2b Draw a labelled sketch of an SEM and explain the important features.

Response

Your sketch should have all the component parts shown in Fig. 1.2a.

The electron optical column consisting of:

a vacuum system – to prevent electrons being scattered by the atmosphere;

an electron gun – the source of the electron beam;

an anode – to accelerate the electrons;

lenses – to focus and define the diameter of the beam;

a detector – to collect information for the image;

a video amplifier – to boost the small detector signal before it is applied to the CRT brightness control;

a CRT display screen – on which to present the image;

deflection coils – to scan the beams over the specimen and the CRT screen;

a magnification control – to limit the area of the specimen scanned by the beam;

a scan generator – to drive the deflection coils in the column and the CRT display.

If you omitted any parts, it would be worthwhile revising this section, before continuing.

****

> **SAQ 1.3a** List the important interactions that occur when an electron beam strikes the specimen and explain the conditions under which these occur. Which of the type(s) you have listed are of use in the SEM?

Response

You ought to have listed *all* the products shown in Fig. 1.3a ie:

> secondary electrons;
> backscattered electrons;
> transmitted electrons;
> Auger electrons;
> visible photons (cathodoluminescence);
> X-rays (characteristic and continuum);
> phonons (heat).

Secondary and backscattered electrons are important for SEM imaging, and X-rays for microanalysis. If you left out any of these you should revise this section before going on.

****

Open Learning   91

SAQ 1.5a — Define the term resolution, explain how magnification is dependent upon it, and list the operating conditions necessary to achieve maximum resolution.

Response

Resolution is the smallest distance between two adjacent points, which can be seen as separate, by the optical system.

Excessively large magnifications do not provide more detail. For example if the SEM can resolve 10 nm and the unaided eye can resolve 0.2 mm there is little point in magnifying more than about 20 000 times.

Highest resolution is obtained in the SEM by using the:

> highest accelerating voltage;
> shortest working distance;
> smallest beam (spot) diameter;
> largest aperture.

****

SAQ 2.2a — Explain the principles of freeze-drying and critical point drying. Discuss the relative advantages and disadvantages of the two methods.

Response

Both techniques allow specimens to be dried, without damage, by avoiding transition through the vapour/liquid interface.

Freeze drying relies on reduced temperature and pressure, to sublime the liquid phase (usually water) to the vapour phase without the liquid phase being present.

Critical point drying uses increased temperature and pressure to convert the liquid to vapour above its critical point, where vapour and liquid have equal densities, so the phase boundary ceases to exist.

Freeze drying is very slow in comparison to critical point drying. Its only real advantage is that it does allow hydrated specimens to be dried, without the use of organic solvents.

****

SAQ 2.3a Explain the principles of vacuum evaporation and sputter coating techniques. List their relative advantages and disadvantages.

Response

Vacuum evaporation involves vaporising metal or carbon from a radiant heat source, in a high vacuum. The minute metal or carbon particles settle on any surface which is exposed directly to the source. Sputter coating is carried out in an atmosphere of argon, below atmospheric pressure. Applying a high negative voltage to a metal target causes ionisation of the gas, and the resulting ions bombard the target. Minute particles of metal are dislodged, and these behave like a metal smoke, settling on contact with any surface.

Open Learning 93

Vacuum techniques, although capable of producing very thin coatings, are very directional. Sputter coating produces much more uniform coatings. At present carbon can only be deposited by vacuum evaporation.

****

| SAQ 2.3b | Explain what is meant by 'coating artefact'. Explain how they arise and how they may be avoided or minimised. |

Response

Coating artefacts are either surface damage or deformities, or deposited structures (agglomerations of particles) and they are a direct result of the specimen coating process. They are often the result of heat damage, which can be reduced by keeping the specimen/source distance as large as possible, and/or by cooling the specimens during the process. Ion or electron etching can be reduced by using magnetic deflectors or by ion beam sputtering.

****

| SAQ 3.2a | Using the values of the constants, h and c determine the numerical relationship between the wavelength, λ (expressed in nm) and the energy E (expressed in J and in kev) in the equation $$\lambda = hc/E$$ $h = 6.626 \times 10^{-34}$ J s; $c = 2.998 \times 10^8$ m s^{-1}; 1 eV $\equiv 1.602 \times 10^{-19}$ J |

Response

It is most important to get the units correct. If we substitute, in terms of units we have

$$(\lambda/m) = \frac{(h/\text{J s}) \times (c/\text{m s}^{-1})}{E}$$

To satisfy the equation E must have the units of J

$$\lambda = \frac{6.626 \times 10^{-34} \times 2.998 \times 10^{8} \text{ m}}{(E/\text{J})}$$

$$= \frac{1.986 \times 10^{-25} \text{ m}}{(E/\text{J})}$$

$$= \frac{1.986 \times 10^{-16} \text{ nm}}{(E/\text{J})}$$

When E is expressed in eV

$$\lambda = \frac{1.986 \times 10^{-25}}{1.602 \times 10^{-19}(E/\text{eV})}$$

$$= \frac{1.2398 \times 10^{-6} \text{ m}}{(E/\text{eV})}$$

and when E is expressed in keV

$$\lambda = \frac{1.2398 \times 10^{-6} \times 10^{-3} \text{ m}}{(E/\text{keV})}$$

$$= 1.2398 \text{ nm}/(E/\text{keV})$$

****

SAQ 3.2b For an X-ray of wavelength 2.5×10^{-10} m, calculate the quantum energy E (in keV) and the frequency of the radiation.

Response

(i) From the equation

$$\lambda/\text{nm} = hc/E = 1.2398/E$$

we can calculate E (remember to use SI units)

$$\lambda = 2.5 \times 10^{-10} \text{ m}$$
$$= 2.5 \times 10^{-1} \text{ nm}$$

$$\therefore 25 \times 10^{-1} = \frac{1.2398}{E}$$

$$E = \frac{1.2398}{0.25}$$

$$= 4.96 \text{ keV}$$

$$1 \text{ J} = 1.602 \times 10^{-19} \text{ eV}$$

$$\therefore E = \frac{4.96 \times 1000}{1.602 \times 10^{-19}}$$

$$= 3.09 \times 10^{22} \text{ J}$$

The frequency of the radiation $\nu = c/\lambda$

$$\therefore \nu = \frac{2.998 \times 10^8 \text{ ms}^{-1}}{2.5 \times 10^{-10} \text{ m}}$$

$$= 1.20 \times 10^{18} \text{ s}^{-1}$$

$$= 1.2 \times 10^{18} \text{ Hz}$$

****

SAQ 3.4a Explain the principles underlying the detection of X-rays and list the advantages and disadvantages of WDS and EDX systems.

Response

WDS systems operate by first filtering X-rays into a particular wavelength (using diffracting crystals). These are them counted using a gas flow proportional detector. Elements are identified from their characteristic wavelengths, which can be determined from the crystal spacing and angle. Quantitation can be made from the count rate.

EDX systems use a cryo-cooled SiLi crystal to collect most of the X-rays emitted by the specimen. A multichannel analyser then counts the number of X-rays occuring at different energies. Elements are identified from their characteristic energies, and their concentration can be derived from the count rate.

WDS offers better resolution than EDX, with less peak overlapping. EDX analyses from smaller areas than WDS, and does not require such large beam currents.

WDS can usually only analyse for one element at a time, EDX can detect almost all elements simultaneously.

WDS can detect elements of atomic number 4 and above, EDX detectors with Be windows can only detect elements with atomic numbers >10.

EDX detectors are more efficient than WDS so analysis times are shorter.

EDX detectors produce spectral artefacts (eg peak overlaps and escape peaks).

Units of Measurement

For historic reasons a number of different units of measurement have evolved to express quantity of the same thing. In the 1960s, many international scientific bodies recommended the standardisation of names and symbols and the adoption universally of a coherent set of units—the SI units (Système Internationale d'Unités)—based on the definition of five basic units: metre (m); kilogram (kg); second (s); ampere (A); mole (mol); and candela (cd).

The earlier literature references and some of the older text books, naturally use the older units. Even now many practicing scientists have not adopted the SI unit as their working unit. It is therefore necessary to know of the older units and be able to interconvert with SI units.

In this series of texts SI units are used as standard practice. However in areas of activity where their use has not become general practice, eg biologically based laboratories, the earlier defined units are used. This is explained in the study guide to each unit.

Table 1 shows some symbols and abbreviations commonly used in analytical chemistry; Table 5 is a glossary of abbreviations used in this particular text. Table 2 shows some of the alternative methods for expressing the values of physical quantities and the relationship to the value in SI units.

More details and definition of other units may be found in the *Manual of Symbols and Terminology for Physicochemical Quantities and Units*, Whiffen, 1979, Pergamon Press.

Table 1 *Symbols and Abbreviations Commonly used in Analytical Chemistry*

Å	Angstrom
$A_r(X)$	relative atomic mass of X
A	ampere
E or U	energy
G	Gibbs free energy (function)
H	enthalpy
J	joule
K	kelvin ($273.15 + t\,°C$)
K	equilibrium constant (with subscripts p, c, therm etc.)
K_a, K_b	acid and base ionisation constants
$M_r(X)$	relative molecular mass of X
N	newton (SI unit of force)
P	total pressure
s	standard deviation
T	temperature/K
V	volume
V	volt ($J\,A^{-1}\,s^{-1}$)
$a, a(A)$	activity, activity of A
c	concentration/ mol dm^{-3}
e	electron
g	gramme
i	current
s	second
t	temperature / °C
bp	boiling point
fp	freezing point
mp	melting point
≈	approximately equal to
<	less than
>	greater than
e, $\exp(x)$	exponential of x
$\ln x$	natural logarithm of x; $\ln x = 2.303 \log x$
$\log x$	common logarithm of x to base 10

Table 2 *Alternative Methods of Expressing Various Physical Quantities*

1. **Mass (SI unit : kg)**

   $g = 10^{-3}$ kg
   $mg = 10^{-3}$ g $= 10^{-6}$ kg
   $\mu g = 10^{-6}$ g $= 10^{-9}$ kg

2. **Length (SI unit : m)**

   $cm = 10^{-2}$ m
   $\text{Å} = 10^{-10}$ m
   $nm = 10^{-9}$ m $= 10$ Å
   $pm = 10^{-12}$ m $= 10^{-2}$ Å

3. **Volume (SI unit : m^3)**

   $l = dm^3 = 10^{-3}$ m^3
   $ml = cm^3 = 10^{-6}$ m^3
   $\mu l = 10^{-3}$ cm^3

4. **Concentration (SI units : mol m^{-3})**

   $M = $ mol l^{-1} $= $ mol dm^{-3} $= 10^3$ mol m^{-3}
   mg l^{-1} $= \mu$g cm^{-3} $= $ ppm $= 10^{-3}$ g dm^{-3}
   μg g^{-1} $= $ ppm $= 10^{-6}$ g g^{-1}
   ng cm^{-3} $= 10^{-6}$ g dm^{-3}
   ng dm^{-3} $= $ pg cm^{-3}
   pg g^{-1} $= $ ppb $= 10^{-12}$ g g^{-1}
   mg% $= 10^{-2}$ g dm^{-3}
   μg% $= 10^{-5}$ g dm^{-3}

5. **Pressure (SI unit : N m^{-2} $= $ kg m^{-1} s^{-2})**

   Pa $= $ Nm^{-2}
   atmos $= 101\ 325$ N m^{-2}
   bar $= 10^5$ N m^{-2}
   torr $= $ mmHg $= 133.322$ N m^{-2}

6. **Energy (SI unit : J $= $ kg m^2 s^{-2})**

   cal $= 4.184$ J
   erg $= 10^{-7}$ J
   eV $= 1.602 \times 10^{-19}$ J

Table 3 *Prefixes for SI Units*

Fraction	Prefix	Symbol
10^{-1}	deci	d
10^{-2}	centi	c
10^{-3}	milli	m
10^{-6}	micro	μ
10^{-9}	nano	n
10^{-12}	pico	p
10^{-15}	femto	f
10^{-18}	atto	a

Multiple	Prefix	Symbol
10	deka	da
10^2	hecto	h
10^3	kilo	k
10^6	mega	M
10^9	giga	G
10^{12}	tera	T
10^{15}	peta	P
10^{18}	exa	E

Table 4 *Recommended Values of Physical Constants*

Physical constant	Symbol	Value
acceleration due to gravity	g	9.81 m s^{-2}
Avogadro constant	N_A	$6.022\ 05 \times 10^{23}$ mol^{-1}
Boltzmann constant	k	$1.380\ 66 \times 10^{-23}$ J K^{-1}
charge to mass ratio	e/m	$1.758\ 796 \times 10^{11}$ C kg^{-1}
electronic charge	e	$1.602\ 19 \times 10^{-19}$ C
Faraday constant	F	$9.648\ 46 \times 10^{4}$ C mol^{-1}
gas constant	R	8.314 J K^{-1} mol^{-1}
'ice-point' temperature	T_{ice}	273.150 K exactly
molar volume of ideal gas (stp)	V_m	$2.241\ 38 \times 10^{-2}$ m^{3} mol^{-1}
permittivity of a vacuum	ϵ_0	$8.854\ 188 \times 10^{-12}$ kg^{-1} m^{-3} s^{4} A^{2} (F m^{-1})
Planck constant	h	$6.626\ 2 \times 10^{-34}$ J s
standard atmosphere pressure	p	$101\ 325$ N m^{-2} exactly
atomic mass unit	m_u	$1.660\ 566 \times 10^{-27}$ kg
speed of light in a vacuum	c	$2.997\ 925 \times 10^{8}$ m s^{-1}

Table 5 *Glossary and Abbreviations used in Scanning Electron Microscopy and X-Ray Microanalysis*

AES	Auger electron spectroscopy
bse	backscattered electron(s)
CRT	cathode-ray tube
cpd	critical point drying
EDX	energy dispersive X-ray (microanalysis)
EELS	electron energy loss spectroscopy
ESCA	electron spectroscopy for chemical analysis
ISS	ion scattering spectroscopy
PM	photo-multiplier (tube)
SAM	scanning Auger microprobe
se	secondary electrons
SEM	scanning electron microscope
SIMS	secondary ion mass spectroscopy
TEM	transmission electron microscope
WDS	wavelength dispersive spectrometry
ZAF	atomic number (Z), absorption (A) and fluorescence (F)
XPS	X-ray photoelectron spectroscopy

References

Echlin, P., Ralph, B., Weibel, E. (eds.) (1978) *Low Temperature Biological Microscopy and Microanalysis*, Royal Microscopical Society, Oxford.

Everhart, T.E. and Thornley, R.F.M. (1960). *J. Sci. Inst.*, **37**, 246, (III).

Fitzgerald, R., Keil, K. and Heinrich, K.F.J. (1968). *Science*, **159**, 528.

Hall, T.A. and Gupta, B.L (1974) in *Microprobe Analysis as applied to cells and tissues* (ed: Hall, T.A Echlin, P. and Kautmann, R.) p. 147–58. Academic Press.

Hendee, C.F., Fine, C.S., and Brown, W.D. (1956) *Rev. Sci. Instr.*, **27**, 531.

Johnson, G.G. and White, W.(1970), *X-ray wavelengths and keV tables for nondiffractive analysis*, American Society for Testing and Materials, Philadelphia.

Murphey, J.A. (1978), *SEM/1978/II*, SEM Inc., AMF O'Hare, Illinois, p. 175.

Murphey, J.A. (1980), *SEM/1980/I*, SEM Inc., AMF O'Hare, Illinois, p. 209.

Pawley, J.B. (1974), *SEM/1974/I*, IIT Research Institute, Chicago, Illinois, p. 27.

Rick, R., Dorge, A. and Thurau, K. (1979), *Quantitative analysis of electrolytes in frozen dried sections. J. Microsc.*, **125**, 239–247.

Robinson,V.N.E. (1980), *Scanning 3*, 15. Wells, O.C. (1977) SEM/1977/I, IIT Research Institute, Chicago, Illinois, p. 747.

/543.0812L417S>C1/

DATE DUE

DEC 31 1988		
JUN 10 1992		